CHICKEN COOPS

Top 50 Chicken Coop Designs for Raising Healthy Poultry

CATHERINE XAVIER

© **Copyright 2020 - All rights reserved.**

The contents of this book may not be reproduced, duplicated, or transmitted without direct written permission from the author.

Under no circumstances will any legal responsibility or blame be held against the publisher for any reparation, damages, or monetary loss due to the information herein, either directly or indirectly.

<u>Legal Notice:</u>

This book is copyright protected. This is only for personal use. You cannot amend, distribute, sell, use, quote or paraphrase any part or the content within this book without the consent of the author.

<u>Disclaimer Notice:</u>

Please note the information contained within this document is for educational and entertainment purposes only. Every attempt has been made to provide accurate, up to date and reliable, complete information. No warranties of any kind are expressed or implied. Readers acknowledge that the author is not engaging in the rendering of legal, financial, medical, or professional advice. The content of this book has been derived from various sources. Please consult a licensed professional before attempting any techniques outlined in this book.

By reading this document, the reader agrees that under no circumstances are the author responsible for any losses, direct or indirect, which are incurred as a result of the use of information contained within this document, including, but not limited to, —errors, omissions, or inaccuracies.

Table of Contents

Introduction

Gallus gallus domesticus, the domesticated chicken, provides food in the form of meat and eggs to human beings. Chickens are an immense source of protein as reported by the Food and Agriculture Organization of the United States in 2007. The history of the domestication of chicken can be traced in the South East and East Asia as a source of entertainment, taking the form of cockfights. Chickens have also been used in conducting scientific research. However, it has not been possible to identify the beginning of the domestication of chickens exactly.

In the present day, human beings are still domesticating chickens even in their backyards, and over time the chicken home has undergone numerous improvements, which now leads us to this discussion on the numerous chicken coop designs. These designs vary greatly to meet the numerous needs of farmers. Join me as we take a look at all of the ideas and options that are available to you.

Chapter One

The History of the
Domestication for Chickens

The history of the domestication of chickens is useful in understanding the best type of chicken coop for a farmer. The annals will also detail the challenges that previous poultry farmers have faced and their solutions for the modern-day.

The source of this history shall be drawn from the United States National Chicken Council, which is non- profit organization acting as the advocate for the United States broiler chicken interests in Washington. The organization has been in existence since 1954 and was at the time referred to as the National Broiler Council but later changed its name to the National Chicken Council in 1998.

The 1800s to the Early 1990s

In the 1800s and early 1990s, there was poultry farming, but it was mostly backyard farming as a means of providing food to the household. This has evolved in the modern-day since chickens are even being kept as pets. The chickens would mostly be reared for their eggs. Chicken meat was not consumed a lot except during special holidays and Sundays.

Upon entering the early 1900s, backyard poultry farmers began to turn poultry farming into a business by selling the chickens to other people who didn't rear them as a means of generating funds for the household. Due to the absence of information at the time, people did not understand the importance of Vitamin D for chickens and how sunlight and dullness contributed to egg production. Therefore, the production rate was much slower than the modern-day.

The 1920s to the 1930s

The egg industry at this time was far much ahead than the poultry meat industry. The meat industry began to formulate by the creation of the broiler chicken, which was being raised solely for meat. This broiler industry commenced in Delmarva Peninsula, Georgia, Arkansas, and New England. People began to understand the requirements of rearing a healthy chicken so that poultry farming would expand to other locations. The crucial factors needed at the time were as follows:

- Space for the chickens to move around, meaning that people needed to purchase land or use their backyard if it was spacious.

- The availability of clean drinking water for the chickens to ensure that they were hydrated to yield better products.

- Poultry feed, which consisted of soybean and corn so that the chickens were healthy and resulting in a nutritious meat by-product for human beings.

- The most conducive weather to rear chickens was also integral; for example, chickens do not like the rain, which was a factor that had to be considered.

Based on historical reports, the discoverer of commercialization of broiler chicken is known as Mrs. Wilmer Steele, who resided in Sussex County, which is in Delaware. This pioneer is reported to have reared 500 chickens in 1923, which she sold for meat to the people. The business was doing very well financially till she managed to own a chicken coop that could accommodate 10,000 chickens, which she built.

The 1940s to the 1960s

The poultry farming industry was not integrated by this time. The hatcheries, farms, feed mills, and processors all acted in their capacity. The hatcheries were the starting point of any intending farmer because it was where farmers would purchase the chicks that he would rear on their farm. The poultry farmers needed the hatcheries because the chicken egg-laying process is technical and not all farmers could do it. Since the hatcheries were crucial, the poultry farmers needed them to manage the poultry farming business from production, processing, and marketing. To ensure that the hatcheries secured their share in the poultry farming production and marketing, they performed these duties.

The feed mills also needed the poultry farmers to purchase their chicken feed. The feed mill entity would even loan poultry farmers money to purchase chicks from the hatcheries which they would feed and sell to the market. Upon the sale of the chickens, the farmers would then pay the feed mills for the purchase of the chicks and the chicken feed.

During this time, after the chickens were slaughtered, the feathers and blood were removed and the chickens sold to market. This type of sale is referred to as 'New York dressed.' This was later developed in 1942 when the government issued approval to a plant in Illinois for the disembowelment of the slaughtered chicken to package it and store it in ice for preservation. This was to make it easier for the consumer to purchase and cook, which is still happening in modern-day.

Due to the development of the broiler chicken market, there was a need for consumer protection. This was initiated in 1949 by the United States Department of Agriculture through a strategy that standardized the quality of broiler chicken going into the market.

In 1952, the broiler chicken became far much more popular than the regular chickens grown in farms by becoming the chicken meat that is most consumed in the United States.

The chicken broiler market became very popular, and a lot of farmers could not handle the financial burden of rearing chickens, especially where they were doing it for the sole purpose of selling them. Therefore, a lot of farmers begun to sell their farms.

The 1940s was the beginning of the integration of the poultry farming entities. Individual companies would handle the production, processing, and marketing of the poultry farming industry. During the mid-1960s, nearly 100% of the chicken broiler industry was being run by companies that had intergraded different poultry farming entities, which are production, processing, and marketing.

This form of integration is referred to as vertical integration to which these companies began to develop even more over time due to the discovery of better medicine, information, and technology, thus securing their hold in the market.

In the year 1959, there was a compulsory legal inspection of the chicken broilers that were being produced. Since there was already accepted by many poultry farming processing plants to have their products inspected in 1926 by the United States Department of Agriculture, this federal inspection was a smooth transition.

By the late 1960s and early 1970s, broiler chicken companies began to market their product through an advertisement on the televisions and the newspapers, thus introducing their brand to the world, which is still happening in the modern-day.

Post-1970s

The broiler chicken industry had developed to this modern-day by the mid-1970s. There was a lot of research and discoveries made from the nutritional aspect of the chicken feed and the benefits of the consumption of chickens by human beings. There was development in pharmaceuticals for chickens, the best breeding practices, and the technology for the production, processing, and marketing of the chicken by-products.

Due to these advancements and federal inspection of the broiler chicken, customers began to prefer the processed chickens to the farm chicken by the early 1980s. This preference grew to the point that chicken meat consumption was at a higher rate than pork in 1985 and beef in 1992.

The broiler chicken market grew so much that in 1991, the first chicken broiler exportation happened from the United States to the Soviet Union. The United States government exported chicken leg quarters, and the Russians began to refer to that chicken piece as the 'Bush Legs' after the then United States President, his excellency George Bush.

The exportation of chicken to Russia and other countries was at a sky-high by the year 2001. Chicken exportation at the time was reported to take up 20% of the entire United States of America production, which was financially quantified to be worth more than 2 billion dollars.

Consumer protection was still in high demand, and as such, the United States Development Agriculture implemented a strategy that there should be a Hazard Analysis and Critical Control Point in the large-scale slaughterhouses. This was done on the 26th of January

1998 with the sole purpose of ensuring that the broiler chicken was not hazardous for consumption due to the chemicals being used on the meat.

With all these developments, the broiler chicken of today is cheaper for the consumer and bigger than the past years, and it will only get better over time, taking consumer protection into consideration.

Based on the above history, it is clear that modern-day has made it much easier for farmers to start their chicken production, and to do so, there is a need for an adequate chicken coop that will meet their needs. A farmer also clearly has to ensure that the chickens are taken care of to have better yield and ensure consumer protection.

An Analysis of Egg Production Over the Years
There has been a lot of concern over the accommodation for chickens over the years, particularly when it comes to layers. Layers are chickens that are reared by farmers with the sole purpose of laying eggs.

The Housing of Layers
There was a major outcry in Europe as to the housing of chickens because they were being placed in cage-like structures. There was a huge demand that alternative housing structures should be put in place to get rid of conventional cages. Due to this major outcry, the European Union was forced to prohibit conventional housing cages for chicken in 1999. There was were even funding being from the national government and animal welfare organizations to develop other housing alternatives.

Over time some chicken coops were found to be efficient but required further improvements while others were considered to be ineffective, and there was no further research conducted on them.

The European Union currently only permits chicken housing systems that are not conventional cages but only enriched cages. There are 2 types of non-cage chicken housing systems, and they are referred to as the floor and aviaries system, which can accommodate thousands of chickens.

The floor system is characterized by placing the chickens on the structure's floor rather than in a cage. There are nesting boxes within the structure to provide the conducive conditions for the chickens to lay their eggs, which have been systematically placed to allow for the egg collection.

The aviaries housing system has a floor system but also a top-level for the chickens to go above. As such, it is often referred to as the multi-tier system. Litter is placed on the bottom floor, and the top-level is organized so that the chicken droppings do not fall on the chickens that are on the ground floor. This system seems to be more economical than the floor system because it can accommodate more chickens due to the usage of the top layer of the building.

It is recommended that chickens should have access to the outside for them to roam amongst each other. Chickens could also be reared on free-range farming where the animals can roam during the day without being in an enclosure, but there should be a house for them to go in, especially at night, to protect them from predators and harsh weather. This type of housing is for the small-scale farmer, and the United States refers to this housing system as a pastured system. The house in place for free-range poultry farming should be portable so that the relocation protects the chickens from parasites in the soil that may be in one area.

There is also a modified cage housing system that is different from conventional cages. There are adjustments inside that have allowed for nesting boxes for the chickens to lay their eggs comfortably, perches for the chickens to rest and substances put in place for the chickens to forage.

During the 1980s, the first paradigm was developed by the Germans, which were equipped with perches for the chickens to sleep and were also different from conventional cages because they could accommodate a large flock. The challenge with this prototype was that there was cannibalism amongst the chickens.

The Edinburgh enriched cage was among the developments that followed in 1998. This housing system did not have the commercial aspect in mind but instead placed the welfare of the chickens as the priority. This is because the enriched cage could only accommodate four chickens with one perch for resting, one nesting box for laying their eggs, and a dust bath that was topped up with litter.

These developments are what led to the modern-day chicken coops that have nesting, foraging, and dusting places and perches to promote the overall happiness of the chickens.

The European Union developed classifications for the modified chicken which were as follows:

- A maximum of 15 chickens in a chicken cage was classified as a small size.

- A range of between 15 and 60 chicken in a chicken cage was classified as a medium size.

- A range of 30 to 60 chickens in a chicken cage was referred to as a large size.

The group size system of these modified chicken cages that will be most accepted will be the Enriched Colony System, which can accommodate 20 to 60 chickens, which is the European Union permissible stocking limit. The chicken cages are spacious for the chickens to move around. There is a nesting space that has a mat made out of Astroturf with nesting curtains, so create a conducive environment for the chickens to lay their eggs. There are also perches and the chickens to rest on. A dusting and foraging area is also available, which is has a pad made out of Astroturf where chicken feed that is placed inside so that the chickens feel the urge to scratch and peck in it. There is also a box here; the excess litter is placed.

As mentioned earlier, there is a ban on the use of conventional cages; however, there are still 15 countries in the European Union that primarily place their chickens in conventional cages such as Spain, Greece, Italy, Hungary, among others. Although these countries aren't cooperating, others are such as Switzerland and Austria, where conventional cages were banned in 1992 and 2010 respectively. The egg production system in these countries is not through conventional cages. Sweden also followed through and banned conventional cages in 2004, and as such, the egg production system that involves caged is only done by way of modified cages. There is a possibility that a majority of the eggs produced within the European Union are by way of modified cages, but there is no accurate statistical information on that subject.

One challenge with the enriched cage system is that the labeling of the eggs after production has not caught up with the developments. This means that the eggs that are produced from modified eggs are still being labeled as caged eggs, so a consumer cannot distinguish as to whether the eggs are from a conventional caging system or a modified caged system despite enriched cages being legally permissible in 2012. This lack of distinction shows the consumers that there is no superiority in the eggs from conventional cages and enriched cages, thus discouraging egg production producers in the European Union from implementing the ban of conventional cages. Another reason for the lack of implementation is that there are still complaints from some animal welfare organizations that, despite the enriched cage system having the additional equipment, the chickens are still being confined. Actually, in 2004 an associate of Eurogroup for Animals known as David Wilkins said, "Let me make one thing very clear. We in Eurogroup and I think I can speak on behalf of the whole animal welfare movement in Europe, are opposed to hens in cages…if the legislative debate is reopened, then we may seek to get enriched cages prohibited completely" Therefore, these European Union countries do not want to commit to enriched cages, thereby making the future of modified cages uncertain. However, these few concerns have significantly decreased when enriched cages were found to preserve the welfare of the chickens by the LayWel Project. The European Union made enriched cages legal and passed legislation to regulate their use in 2012. This passed legislation gave animal welfare organizations such as Compassion in World Farming and Royal Society for Prevention of Cruelty to Animals a duty to push individual states that have not outlawed conventional cages.

The Challenges Facing Egg Production

• *A Decrease in the Economy in the Aspect of Egg Production*

In 2004, a report was given to the European Commission, which stated that the ban of conventional cages could be very expensive to the 15 European countries that have not implemented the law. The stated figure was that it would cost egg producers 160 million pounds to shift from conventional cages floor or multi-tier chicken housing systems. The report also reflected that the cost of production of an enriched cage and the conventional cage is nearly similar, but there was no indication of the entire financial cost of the egg production industry as a whole.

There is also research that the outlaw on conventional cages will result in an increased cost of production. This will make the eggs more expensive for the consumers to purchase by nearly one percent per egg.

The reduction of egg production in Germany from 2000 to 2007 was reported to be at 13%. This arose because Germany completely banned cage systems through very strict legislation. The result of this has not been encouraging to the 15 European Union countries because of how costly it was for the country to make the shift. Another reason is that in Germany, the egg production rate is low, and they are at a point where there is a need to import eggs to meet the needs of the consumer.

• Health and Safety

There was a report in 2005 from the European Food Safety Authority Scientific Panel on Animal Health and Welfare, which established that there was egg contamination in the modified cage systems. The report detailed that the eggs were exposed to bacteria when laid on the nesting boxes as compared to on a wire. This, therefore, meant that the eggs from conventional cage systems were safer to consumers than those from alternative housing systems. Based on this report, the European Commission initiated 2 programs referred to as the Safehouse and Rescape. The projects were initiated in 2006. Safehouse looked into the causes of egg contamination while Rescape was to establish solutions to these egg contaminations, for example, the chicken maintenance and egg packaging, among others.

• Environmental Concerns

The biggest threats to the environment concerning egg production is the ammonia released, the energy consumed, the loss of nitrogen and phosphorous. The greatest concern on this list is the ammonia emissions. This can be illustrated by the fact that in 2008, the Netherlands put in place certain legislations on the limit of ammonia emission when it comes to poultry farming. It has been established that alternative housing systems are more detrimental to the environment than conventional cages. However, the accurate reports on this are yet to be proven. There are still these other environmental concerns such as air pollution from the chickens, water pollution, and the use of land for egg production.

• *Human Exposure to Diseases*

There is a major concern that people who rear chickens or work in poultry farming may be exposed to diseases such as avian influenza. However, this has not entirely been proven. The health and safety that has been proven is in a barn chicken farming system. A report indicated that the workers from a barn system come into contact with more dust as compared to those working in conventional cages. This level of dust exposure surpassed the limit of the United Kingdom. The workers from the barn chicken housing system were also exposed to far more ammonia than those from cage housing systems. However, the exposure was high for a minimal amount of time but not for the entire day. Thus, those that work in barn housing systems should ideally wear respiratory protection such as masks because, due to the design of the structure, one simply cannot increase ventilation.

Chapter Two

The Rationale Behind Owning Chickens

As earlier discussed, chickens were being reared by human beings in the past for various reasons, and the benefits of owning chickens have significantly improved over time. Below are of a few reasons as to why people should strive to house chickens in their backyard.

Chickens are a Source of Information

There are people that rear chickens to learn more about them. In the modern-day, it is very rare for someone to be going to work and happen to come across a chicken in the streets. Many people consume this animal yet have never seen one with their eyes. Our dear children grow up in a world where they may never even see a chicken except when it is on their play or as a picture or drawing in their books. Therefore, some parents rear chickens in their backyard to show their children the practicality of life as a means of educating them on how the chicken food products land on their plate and the importance of preserving such an animal. It also helps children understand the cycle of life in that the chicken is a source of food that should be mercifully slaughtered so that a human being can consume to gain healthy nutrients, thereby showing the need to ensure that the chickens reproduce as well. Children also learn how to take care of animals, which in turn contributes positively to their upbringing.

Waste Depletion

Chickens are great to have in your backyard because it ensures that there is no wastage of food in the household. This means that the left-overs that would go into the trash are instead given to the chickens, which would satisfy them more, especially where it is a small flock. The variety of the food being given to the chickens provides multiple nutrients, which in turn leads to healthier by-products from the chickens. Therefore, owning chickens in a household ensures that there is no money spent on food that goes to the trash but instead creates an effecting food recycling process.

Chickens are a Source of Fertilizer

Chickens consume food, and as such, they have to excrete the food. This may discourage some people from rearing chickens, but these chicken droppings are also beneficial to human beings because they can be a great source of fertilizer. The collection of chicken droppings is also not difficult because a majority of chicken coop designs have removable trays that collect the dropping, and the farmer simply needs to pull out the tray, and all the droppings will be collected at once. The fertilizer will be nutritious for the crops that one grows or intends to grow. Most people in the United States have but chicken fertilizer, but if one owns chicken, they could be developing the manure they need in their backyard. However, there is a process of forming this chicken manure; one simply doesn't take the fresh dropping and put in their garden. The chicken droppings should be placed in compost bin so that over time it shall become very nutritious for the plants and also because placing the droppings in the garden immediately will result in a lot of nitrogen being released to the plants which would result in fertilizer burn on the plants. Upon making the chicken manure, one moderately places the manure between the rows in the garden, which will, in turn, be infused into the soul from sprinkling water or rainfall.

Reduction of Insects

Although the chickens will be fed, they are animals and will always search for food elsewhere as well. They do so by scratching the ground where all kinds of insects live, such as ants, worms, cockroaches, amongst others which chickens consume. However, there is a concern that the chickens could get ill from consuming some insects; it would be difficult to stop them from doing so, especially if the chicken coop is one that a chicken can access the ground, but it is not common. The advantage of chickens consuming insects is that the insects will decrease in the compound.

Chicken Rearing is Recreational

Over time, chickens will become a source of entertainment. It is very relaxing watching chickens play amongst themselves. There are even people who house chickens like pets because they being to bond with the owners. Most farmers enjoy the time when they go to feed their chickens as well. This is seen where chickens will begin to run towards their owner in the hopes of being given food and even in some cases flapping their wings to jump on the owners' lap so that they can be pet.

The Protection of the Welfare of the Chicken

There has been a large outcry regarding different poultry farming companies manage their chickens. There have been reports of torture and cruelty of chickens within the poultry industry. Owning chickens gives one a sense of peace that the by-products from the chickens were not obtained cruelly. In explaining this, this book shall look into the Helensville Poultry Farm report that indicated the violation of chickens' animal rights.

Helensville Poultry Farm Report

There was an article written in 2019 that detailed that this poultry farm murdered their chickens. By December 4th of 2019, there were approximately 190,000 chickens that died from poor poultry management. The ventilation system of the chicken house was run by electricity, and there was a power outage, but the company didn't take measures to protect the chickens, and as a result, the chickens died from suffocation. There was no oxygen supply in the chicken house; therefore, the chickens couldn't breathe. The workers didn't ensure that there was a consistent observation of the chicken to the point that they were found dead in the morning. Although the poultry farm didn't intend for this to happen, the incident was still avoidable, and the chickens died slowly, which could be considered to be a form of torture.

In New Zealand, there is a company referred to as Tegal, which is the primary chicken supplier within the area. Tegal sources its chickens from Helensville Poultry Farm, which caused its chickens to die from suffocation. The concerning aspect is that Tegal assures its customers that the chickens from the company are well taken care of, yet that doesn't seem to be the case. This would be considered to be a misrepresentation to the consumers of the product.

A Better Food Quality

Chicken by-products can be purchased from various places such as malls, restaurants, amongst others. However, the consumer of this food has no idea of the contents of the eggs or meat they are consuming because some of the products have preservatives and chemicals which could be harmful to the health of a human being over time. The consumer is also not aware of how the chickens were managed; for example, if they were tortured or not. By rearing chickens in your backyard, all these gaps will be filled because, as a poultry farmer, one will ensure that the chickens eat the quality feed and are happy so that the by-product is of high quality.

Nutritional Benefits of Chicken Meat and Eggs

Poultry meat and eggs are a great source of protein and other minerals and vitamins such as Vitamin A, E, and B12. In the past 20 years, eggs have been proven to have immense nutritional benefits. The ability to include other nutrients in eggs through the chicken feed for chickens is of great importance to the human body. This means that despite eggs being a great source of protein, one could even go further and add other nutrients such as folic acid, omega- 3 fatty acids, among others.

The National Nutrition Monitoring Bureau reported that the Indian population had a lot of nutritional deficiencies that were being reflected among the people. The report indicated that nutritional deficiency and the lack of energy amongst the children were extremely concerning. This nutrition deficiency rate was even higher than in the 1900s. The children from rural areas were not having enough protein and other nutrients in their diet, thus contributing to stagnated growth and poor mental development. The pregnant and breast-feeding women's dietary composition was also unhealthy for babies. The basic foods in India are mostly carbohydrates, which are wheat and rice. These foods lack amino acids such as lysine and threonine, which are significantly higher in chicken meat and eggs. Wheat and rice have very minimal protein, which can be extremely detrimental to children and pregnant mothers.

• Chicken Meat is Healthier Than Red Meat

The World Cancer Research Fund and other agencies have reported that the consumption of a high amount of chicken in a week is much healthier than red meat. They have reported that consumption of over five hundred grams of red meat is unhealthy, which isn't the case for chicken meat.

The fat content in chicken is significantly less than that of red meat. Besides, half of the fat content in chicken means is healthy mono-saturated fat, and a very small portion is composed of unhealthy saturated fats. Meanwhile, red meat is mostly composed of unhealthy saturated fat. Trans- fat is one of the causes of heart complications in human beings, which significantly contained in red meat while chicken meat doesn't have any trans-fat.

• The Advantages of Omega 3 Fatty Acids

Chicken meat and egg have a good amount of omega 3 fatty acids. This dietary component has been established to be crucial to the human body because it reduces the chances of getting heart complications, cancer, and asthma. It is also essential in the development of the brain, boosting one's learning capabilities, making it very important in the development of a child. There are 2 types of omega 3 fatty acids, which are n-6 and n-3 fatty acids. N-3 fatty acid is not very common in vegetables, but flax seeds contain a very high amount of it. Because chicken diets are vast, their by-product contains a sufficient amount of N-3 fatty acids. N-6 fatty acids, on the other hand, are very common in plants. The amount of N-3 fatty acid can also be significantly increased in chickens by adding a few flax seeds (this has the highest amount of N-3 fatty acid), fish meal, fish oil, and rapeseed in the chicken feed. However, it wouldn't be advisable to use a lot of the fish products because of the harsh odor that it will give off.

• Selenium in Eggs and Chicken Meat

Selenium is a vital antioxidant for the immunity of a human being. It has also been subject to much research to ensure that there is no formation of cancer in the body. This potent anti-oxidant also affects fertility in men because it affects the sperms' metabolic energy rate, thus affecting its movement. A deficiency in selenium also leads to miscarriages in pregnant women. Low selenium consumption has been proven to lead to heart diseases and even depression. Selenium deficiency has been established as a huge problem due to its absence in the soil where we grow our plants.

Therefore, to ensure that there is an adequate intake of selenium, one should add 0.2 milligrams of selenium into the chicken feed that causing the selenium content of the chickens' by-products such as eggs and met to be significantly higher.

• The Advantages of Folic Acid in Chicken Meat and Eggs

Folic acid is essential during pregnancy. A report in Uttar Pradesh indicated that babies born from mothers with folic acid deficiency would have defective nervous systems. Folic acid can be obtained from vegetables, but after cooking the vegetables, half of the folic acid is destroyed. Eggs and chicken meat have a great amount of folic acid and are, therefore very nutritious for the diet of a human being, especially a pregnant woman. To increase the folic acid content even further, one could supply chicken feed that is folic acid dense, thus increasing the folic acid content even further.

• Advantages of Iodine in Eggs

Iodine is a very vital nutritional compound that is very beneficial to the human body, particularly the thyroid glands. Iodine deficiency has terrible consequences on the human body. A pregnant woman who doesn't consume iodine dense foods such as eggs could give birth to a child that has mental conditions and stillbirths. This is because iodine is crucial during the gestation period. After all, it ensures brain development. An adult with iodine deficiency suffers from a condition known as goiter.

Vegetables also contain iodine, but in some areas, the soil does not contain iodine, which means that the plants will be iodine-deficient as well. This would mean that the human being would consume vegetables that are iodine deficient. The introduction of approximately 5 milligrams of iodine in the chicken feed will significantly increase the iodine content in an egg, which will, in turn, be healthier for human beings.

• Eggs Contain Lutein

Eggs contain lutein, which is mostly found in plant-based foods. This is a carotenoid that has been proven to be very beneficial in the diet of a human being. Lutein has been reported to have anti-inflammatory properties. Lutein is significantly known to be beneficial to the eyes and aid the elderly when it comes to diseases that affect the muscle that could lead to vision complications and even blindness.

The health benefits of lutein are that it reduces the chances of one getting cancer and increases the health of the heart, thus preventing any cardiovascular diseases. It has also been reported to prevent the spread of chronic diseases.

• Cholesterol in Eggs

Chickens need cholesterol in their diet. The reason for this that cholesterol is one of the nutritional components that ensure chickens can hatch eggs, thus increasing the flock. The absence of cholesterol will derail embryo development, causing chickens not to hatch eggs.

A chicken owner can easily have a small amount of control over how much cholesterol is the eggs. This means that where the owner feeds the chickens' different types of feeds, the cholesterol level could decrease by 10%. The introduction of copper into the chicken feed could also decrease cholesterol content by nearly 31 percent. The introduction of garlic into the chicken feed by 8 percent could reduce the cholesterol by 24%.

Upon the consumption of cholesterol into the body, there are means of reducing its rate of absorption. This can by introducing fiber and saponins into the diet. Saponins can be found in vegetables, and they attach the cholesterol to the small intestines, which are later excreted out of the body. The fiber, on the other hand, cleans out the intestines, thus pushing the cholesterol out of the body as well.

• The Recommended Amount of Egg Consumption

Eggs are very nutritious. There have been concerns in the past that the consumption of eggs daily will lead to high cholesterol, which isn't the case. Eggs are also relatively cheap and, therefore, affordable to a majority of the population, but owning the chickens would be better because one would be able to contribute to the nutritional contents of the eggs.

In 1990, there was a report that most developed countries had a low consumption rate of eggs due to the fear of contracting heart diseases because the cholesterol level would clog their coronary artery. An egg that weighs 60 grams has approximately two hundred milligrams of cholesterol, and the recommended daily amount of cholesterol consumption is 300 milligrams. This is why a lot of people were concerned. However, the fat content of an egg decreases the cholesterol absorptions into the bloodstream due to the presence of monounsaturated fatty acids. The British Nutrition

Foundation, together with the Canadian, Australian, and Irish Heart Foundations, has also reported that egg consumption is not associated with heart conditions. However, everything one consumes should always be taken in moderation; for example, one should not eat 10 eggs a day.

Chapter Three

Factors to Consider when Choosing a Chicken Coop Design

Basic Characteristics of a Chicken Coop

In as much as there are vast chicken coop designs, there are certain standards that every chicken coop must-have for various reasons. These include food supply to the chickens, hygiene of the chicken coop, ventilation, among others. These standards must be in every design, so the only question that a farmer should be asking is the degree of its availability. Therefore, here are crucial features for any chicken coop should have:

- Eggs are fragile, and such there should be nesting boxes available to protect them. They should also be dark and private to promote the egg-laying process. The nesting box should also be cushioned with a material such as dried grass to protect the eggs laid.

- A run would ensure that the chickens can roam around and be happy; however, we shall soon see that there are chicken coops that do not have an outside run. A farmer should purchase a chicken coop with a run so that they do not incur an additional financial cost in the attempt to make one.

- A farmer should find it easy to clean the chicken cop and as such a removable roof would be beneficial

- The chicken coop should be able to protect the chickens from predators, for example, by having an elevated floor.

- Chickens need to feed and hydrate, and as such, there ought to be a chicken feeder and water containers available.

- There should be storage to keep the chicken feed. The storage space should be dry to ensure that the chicken feed does not grow mold because it would be a health hazard to the chickens

- In the absence of an outdoor run, a farmer could use chicken wire and wood to make one so that the chickens can move within.

There are also other features that a farmer should consider having which shall be explained below.

The Size of the Chicken Coop

The size of the chicken coop is one of the basic factors to consider when selecting the chicken coop design. The size of the chicken coop will solely depend on the number of chickens a farmer intends to rear. However, the standard is that it should at least have 4 square feet of coop space for each chicken and 10 feet squared for the outdoor run space for each hen. Therefore, to rear 10 chickens, one would need 40 square feet of coop space. The ample space per chicken will protect the hens from health issues spreading amongst them quickly but also allow socialization amongst the chickens.

However, these requirements are not fixed, but a farmer should at least not go below 3 square feet of coop space per chicken.

There must be no overcrowding in the chicken coop because it would be a health hazard for the chickens, especially if one of them gets an infectious disease. Another issue with overcrowding is that the chickens could trample on one another in the process of accessing food, thus killing or hurting each other.

The Elevation of the Chicken Coop

One of the challenges a farmer could face is the possibility of predators attacking the chickens as a source of food. To curb such a challenge, a chicken coop ought to be elevated to ensure that predatorial animals cannot dig underneath to access the chickens inside the chicken coop.

Other than predators, some animals could create a home for themselves inside the chicken coop, for example, rats.

Another advantage of having an elevation in the chicken coop is that it promotes ventilation inside the coop, thus making the chickens even more comfortable.

The Availability of Nesting Boxes

Chickens are raised as a source of meat and eggs. Therefore, where one intends to obtain eggs from chickens, there must be a nesting area inside the chicken coop. This nesting area creates comfort for the chickens to lay their eggs. Some chicken coops come within built nesting areas while others are sold separately. A farmer also has the option of building a nesting box for his flock, which doesn't require a material that is difficult or expensive to acquire.

The Importance of Nesting Boxes

Nesting boxes create a conducive environment for the chickens to lay their eggs because it provides privacy, darkness, and safety. These factors cause other chickens to not play in the nesting box since they will identify it as their egg-laying location. The absence of nesting boxes doesn't mean that the chickens will not lay their eggs; they will, however, the number of eggs laid will be less, and they will be scattered all over the chicken coop. This level of egg exposure in the chicken coop could lead to contamination of the eggs, which would be unhealthy for consumption. The egg is laid on the floor could break as the chickens play with one another and eventually spill in the chicken coop. This will cause the chicken coop to be dirty as well, thus forcing a farmer to clean the chicken coop much more, thus increasing the workload. Nesting boxes also make it easier for the farmer to collect the eggs since they are all together.

The nesting boxes should also be spacious. This will give the chickens enough room to enter the nesting box, lay their eggs, and leave. Therefore, a chicken should be able to fit within the nesting box comfortably. It is, therefore, crucial to identify the size of the chicken breed before selecting the nesting box. Space in the nesting area is an important factor, as it prevents the chickens from standing on the eggs and breaking them. This prevents extra mess in the coop and a lower egg production rate.

It would also be important for the chickens to have access to the nesting box easily. At first, it would be advisable to place the nesting box at the bottom of the chicken coop so that the chickens can identify with them. After a short while, you can then raise the nesting box so, but it should still be at an accessible height, one could even use a ladder for this, especially in a multi-tier chicken coop.

One nesting box should be enough for two to three chickens. However, chickens will end up choosing a nesting box that they like more than the others, so there should be at least be enough for the chickens to choose from.

Nesting Boxes that can be Purchased

As earlier discusses, nesting boxes can be built, but there are a few nesting boxes that are available for purchase such as:

- **Miller Single Chicken Nesting Box**

This nesting box is made out of plastic, thus making it long-lasting. This is because plastic does not corrode or deteriorate. The fact that it is made out of plastic means that it is waterproof and can easily be cleaned without fear of rotting. This nesting box is to be mounted on the wall, and an inclined roof will discourage the chickens from excreting on top or inside it. The roof will also ensure that the chickens will not rest on top of it.

- **Roll Out Nesting Box with Curtain**

This nesting box makes it dark and very private for the chickens, thus creating the conducive conditions for the chickens to lay their eggs. The farmer will also find it easy to collect the eggs.

- **Brower 6 Hole Poultry Nest**

This nesting box is distinct because it is of a large size, which makes it more comfortable for the chickens. The nesting box is made out of galvanized still, thus making it long-lasting. The steel also acts as an insulator, thus adding warmth within the nesting box, which is a contributing factor for the chickens to lay their eggs.

There are various ways that one could make their chicken coop. Below are a few ways:

- Take a five-gallon bucket and a few covers which can be bought if they aren't available. Open up the cover by cutting within it. The diameter of the lid should just be enough for the chickens to comfortably get inside and prevent the eggs from falling out of the bucket.

- One could also build the nesting box using wood. One would first have to build a box that has an opening at the anterior for the chickens to access. A lip should then be installed at the front of the nesting box to prevent the eggs from falling out.

The Maintenance of a Nesting Box

It is still very important to take care of the nesting area. The following are a few of the ways of taking care of the nesting boxes:

- It is important to add some material such as dried leaves in the nesting box so that when the eggs are laid, they are protected from cracking or breaking. This padding will also create some form of comfort for the chickens, thus encouraging them to lay the eggs inside the nesting box and not elsewhere in the chicken coop. There is also a challenge that the chickens tend to push the padding out of the nesting box when walking in and out.

- The nesting boxes should be cleaned, especially if a chicken excretes inside it. Regular cleaning would also be important to replace the padding material inside the nesting box.

- There is also the introduction period of the nesting box to the chickens since they will not know what it is. One could train the chickens by placing dummy eggs, which can be purchase or small balls so that the chickens associate them with eggs.

- Insects are also a problem inside the nesting boxes. One could curb this challenge by using nesting material that is an insect repellant or purchasing herbs that discourage insects from going into the nesting area and also releasing a pleasant scent.

The Hygiene Maintenance of the Chicken Coop

It is recommended that a farmer should clean the chicken coop weekly. However, this could vary because of different climatic conditions and the amount of excretion from the chickens. However, to promote hygiene, a farmer should remove the chicken dropping every single day so that by the end of the week, the chicken coop will not be extremely filthy, and it will also be easier to clean by then.

The cleanliness of the chicken coop will also promote the health of the chickens, especially the growing chicks.

During the weekly cleaning, the farmer should clear out all the chicken drooping, replace the bedding, clean using water that has white vinegar in it, and allow air circulation before replacing the bedding.

Since the farmer will be cleaning the chicken coop, it would be important to select a chicken coop that would be easy to clean and not restrictive.

The features that should be available to ensure ease of cleaning should be as follows:

- It should be easy for anyone to go inside the chicken coop. Therefore, the door should be large or one that can be removed.

- There should be more than one way to get inside the chicken coop to reach numerous areas for the purposes of cleaning.

- The roof should be capable of being removed during cleaning.

- It is recommended that the nesting box and perches should be removable, but if they are not, one could use a hose to clean the chicken coop.

There should be the availability of food and water for the chickens.

Chickens need food and water to sustain life. Therefore, a farmer needs to consider where the chicken feed should be stored. The storage area of the chicken feed should be dry to prevent mold from growing inside the feed.

Chicken feed and water are essential for the survival of chickens. The absence of sufficient quality chicken and water is fatal to chickens. The chicken feed and water supply should also be clean, free from any bacteria to ensure that the chickens do not get sick. Dirty water of food can kill the entire flock.

The following are the importance of water for chickens:

- The blood of a chicken is 70% water; therefore, water is very crucial in sustaining the livelihood of a chicken. The failure to supply would result in the death of chickens,

- The absence of water could lead to no egg production. Eggs are also made up of water, and the absence of that would lead to no eggs being produced.

- Water is crucial in helping chickens to feed because saliva is released, thus making it easier for the chicken to swallow the food. The saliva in the body of a chicken is composed of water and enzymes that facilitate the digestion of food. Water, therefore, facilitates the digestion process of a chicken.

- Chicken feed is usually very dry, and the supply of water makes it easier and smoother for the chickens to consume. The absence of water can cause the dry consumed chicken feed to clog the carotid artery, thus leading to the decreased blood supply to the brain, which could lead to the death or paralysis of a chicken.

- The large intestines if the chickens have the sole purpose of absorbing water into the body.

- Water is made up of a mixture of hydrogen and oxygen gas. Oxygen is essential for chickens. The oxygen is delivered to the blood cells so that it can create energy for the functioning of the body.

- During hot days, chickens need water to cool off their body. The increased outside temperature causes them to pant and flap things more as a means of cooling off their body. The absence of water, especially during the hot days, can result in heat stress, which can be fatal to chickens.

Feeding chickens is very important for the following reasons:

- Failure to feed chickens can lead to starvation, which will result in the slow death of the chickens, a form of torture and cruelty.

- Chicken feed supply nutrients that are essential for the growth of a chick, failure to feed them quality feed can result in the stagnation of their growth.

- Chicken feed has nutrients that improve their immunity, thus reducing the chances of the chickens getting diseases.

- Chicken feed nutrients have the essential ingredients for egg production. Poor quality feed can decrease the egg production rate.

- The quality of chicken feed results in healthier means and eggs for human consumption.

- Quality feed ensures that the chickens are healthy and strong.

- Feeding the chickens sufficiently makes them happier and more comfortable.

A farmer could opt to feed chickens inside the chicken coop or outside. There is a debate that it would be safer to feed the chickens outside the chicken coops, but those who wish to feed those chickens have to comply with certain requirements. However, it should be noted that feeding the chickens outside the chicken coop simply means the chickens should be fed inside the chicken run so that they are still safe from predators.

The reasons why feeding the chickens outside the chicken coop would be better are as follows:

- Pouring water into the chicken coop will lead to the growth of mold and fungi on the structure, which wouldn't be a conducive environment for the chickens to live in. Where the chicken coop is made out of wood, it could cause the chicken coop to rot if the wood is not rot-resistant. Metallic chicken coops could also rust in the presence of water. This will, in turn, affect the durability of the chicken coop.

- The wasted chicken feed that fell on the floor could become moldy well because of rotting. Molds release a toxin that could kill the chickens when consumed, and the spores released from the mold be a health hazard.

- It would be more work to clean the chicken coop constantly because of the spillage, but if the chickens feed outside, the spillage will go into the soil instead.

- A chicken coop design that isn't well enclosed could lead to rats going inside due to the filth generated from the spillage of the chicken feed.

- It would be more expensive because of the wasted food inside the chicken coop.

There are still some ways to curb some of the challenges of feeding the chickens inside the chicken coop, and they are as follows:

- The chicken feeder should be raised from the floor to prevent the chicken from scratching the feed in the attempt of spilling it on the floor.

- The lip should be big enough to contain the chicken feed and avoid spillage sufficiently.

- The chicken coop would be better off with nipple drinkers to supply water, thus reducing wastage of water from spillage.

The Type of Chickens being Raised

There are numerous types of chickens, and a farmer will select the type they want based on the primary product they source from the chickens. For example, if a farmer only seeks meat, then they will go for a broiler, but if it is only eggs, then he will seek a layer. Therefore, the chicken coop should be one that can meet the needs of that specific type of chicken. An example would be the lighting within the chicken coop, which is an important factor in the egg production rate of the chickens, especially during winter. Chickens naturally don't lay eggs during the winter; therefore, the farmer needs a chicken coop that can provide light to create a conducive environment for the chicken to lay eggs. It is also important to note that the light shouldn't be availed throughout but for one or 2 hours because a lot of light as well could cause the hen discomfort. The light in the chicken coop would also assist the farmer in seeing properly when cleaning the chicken coop, collecting the eggs, and when providing feed to the hens.

The following are a few of the chicken breeds that can be raised.

• Issa Brown

The name of this chicken coop is a brand name that has been copyrighted. The personality of this chicken breed is that it is very calm. The feathers of this chicken breed have a black and white pattern. A company in France established the breed. The same company placed a patent on the chicken in 1978 because it was very good at laying eggs. Since 1978 to the present day, the chicken breed is still very popular for backyard chicken coop owners and large-scale poultry farmers as well.

• Astralorp

This chicken breed is very popular in Australia. The chicken breed is very beautiful due to its black, white, or blue-colored feathers. The breed also has a high egg production rate and has a very friendly personality.

• Plymouth Rock

This chicken breed is a lovely breed for a chicken owner who has never kept chickens before. The reason for this is because their temperament is calm, and they respond quickly, thus making an owner feel more comfortable. The breed was discovered towards the end of the 19th century in America in a town known as

Plymouth Rock. The breed was very popular because it has a great egg production rate and their relaxed personality

• Barnevelder

This breed is a gift from the Dutch. The breed is among the most popular chicken breeds in Australia due to its unique eggs and appearance. It is a great breed to have on your backyard because it stands out. There are, however, other chicken breeds that are more popular since they have a higher egg production rate, but the features of this breed are so distinct that it still holds its own in the market place.

• Naked Neck

This chicken breed has a unique appearance. The breed has thin and exposed neck i.e.; the neck has no feathers; hence the name, Naked Neck. This breed was discovered in Transylvania, and its neck gives it a resemblance to a turkey. There has been speculation that the naked neck is a crossbreed between a chicken and a turkey, but researchers have proven that this theory is false. This chicken breed would be unique to own in your backyard.

• Orpington

This is the cutest chicken because it is fluffy and would be a beautiful addition to an already existing flock or the first chicken to be owned. The chicken breed is known to be an amazing mother to her chicks. The chicken breed was developed in Britain by poultry farmers towards the end of the 20th century. This would be perfect for an area that is cold because the breed was designed to withstand winter in England but at the same time have an incredible egg production rate.

• **Silkie**

This chicken breed is very small and weighs between 1.5 to 2 kilograms. This breed is very beautiful and very distinct from other breeds such as the ISA and Plymouth Rock, amongst others. This breed leaves a lasting impression due to its beauty and unique appearance. They are such beautiful creatures that Marco Polo couldn't resist taking them with him after he found them in China.

• **New Hampshire Red**

This chicken breed is social, and it produces very scrumptious eggs. The chicken breed is very calm and is a great mother to their chicks. The appearance of the chicken is quite basic, but it makes an excellent flock to have. Poultry farmers in New Hampshire created the breed at the beginning of the 19th century. The need for this development was to improve the Rhode Island Reds further and have a chicken breed that would be named after them. The aim was to have a chicken with a high growth and egg production rate, and capable of withstanding the cold weather. Their development process was a success because, in 1935, New Hampshire was born.

• **Frizzle**

This chicken breed would stand out in a flock of chickens. The feathers of this chicken breed are characterized as appearing as if the feathers have been blow-dried because they are curly outwards, making them very fluffy. The breed makes a lovely companion and would, therefore, make a beautiful addition to your flock or backyard. When purchasing this chicken breed, one has to consider the ventilation system being adjusted to suit their body temperature because they release more heat than other chickens. This chicken breed can be bantam or regular-sized, which would be great to consider when choosing the size of the chicken coop to accommodate them comfortably.

• Belgian d'Uccle

Just as human beings come in different shapes, sizes, and colors, so does this chicken breed. The temperament of this chicken coop is very sweet and lovable. The breed was first discovered in a town known as Uccle in Belgium.

• Cochin

This breed was the beginning of the popularity of chickens for human consumption. As discussed earlier, chickens were not very popular; there were kept to walk around the farms and maybe be consumed by some families during the holidays and Sundays. The sole product that a majority of people consumed from chickens was eggs. However, when human beings began to consume chicken meat more, the Cochin breed was the most popular.

• Rhode Island Reds

This chicken breed is currently very popular. The breed is great for laying eggs, peaceful, and easy to take care of. The breed is known to be very adaptable in different environmental conditions such as snow, rain, and sunshine. Therefore, this chicken breed would be great for a seasonal area and would adjust to the ventilation system of the chicken coop as long as there is at least sufficient air. The chicken breed might appear intimidating, but they are very friendly to the point that they have been proven to be great pets. The breed was established in Rhode Island hence its name towards the end of the 18th century. The popularity of this breed grew so much that it became identified to be the state bird of Rhode Island.

• **Polish**

This chicken breed has a unique appearance, particularly on the head, and is very beautiful. The history of this chicken breed is unknown. They are very friendly chickens and would be a companion and an accessory in the backyard of any chicken owner.

• **Leghorn**

This chicken breed was established in Livorno. The name is Italian, which means Livorno, which is located in Tuscany. The chicken breed is curious, very friendly, and white in color. The chicken is known for being great egg layers as well. The breed was introduced in Britain toward the end of the 18th century but became very popular in America and Australia. There was a lot of interbreeding amongst the chicken breeds at the time, but in the early and 1990s, this breed was the most popular purebred in Australia. The breed is known for laying large brilliant white eggs.

• Sussex

This chicken breed would be perfect as a pet because they are very social chickens and easily adjust to the environment they are placed in. This is a beautiful breed with black and white feathers. The breed enjoys the company of human beings, thus making them great chickens to have in your backyard. The chicken breed is, therefore, therapeutic due to its friendly nature and foraging habit. The breed also has a long family history.

• Araucana

This chicken breed leaves a mark in terms of appearance due to the shape of its feathers. This breed is well known because it lays blue eggs. The breed has an energetic temperament and would be a unique addition to have in the backyard of any chicken owner.

• Wyandotte

This chicken breed is very beautiful with black and yellow patterned feathers. The stunning appearance of this chicken would be an accessory in the backyard of any chicken owner. The chicken breed was developed in North America in the year 1883. Despite the popularity of the Sebright chicken breed, there was a need for another breed that would be more sufficient to the people, especially in size since the Sebright was small. The Wyandotte took some time to hold its own in the market, but when it did, it became very popular because the chicken was great for laying and meat.

• Minorca

The chicken was named after the island on the coast of Spain. There are strong and hyper chickens but also elegant in how they move. This chicken breed has plain black feathers or plain white feathers. In the 1800s, the breed was taken by the British, and they became extremely popular for nearly 100 years. Their popularity dwindled in the late 1990s due to crossbred chickens, which were more reliable egg layers. Their decrease in popularity has placed them near extinction.

• Faverolles

This chicken breed has light brown and white feathers, thus contributing to their beauty. The also very gentle in nature. The breed originated from Northern France and was first introduced to the public towards the end of the 20th century. The breed isn't a pure breed because it is a combination of Houdans, Cochins, and Dorkings that are large in size and healthy. The chickens selected out of the breeds for interbreeding should be healthy and consistent in laying eggs. This, in turn, ensured that the Faverolles would be great layers for poultry farmers. They are also very fluffy and would, therefore, make great pets.

• Sebright

This is chicken breed is very beautiful with black and white patterned feathers. It is an exotic breed that is characterized to be curious and welcoming to people. The breed was created in the early 1800s by a member of the British Parliament, referred to as Sir John Sebright. He was determined to create the perfect chicken in his eyes. During his discovery, he created the Sebright Bantam, which was known for its small size and beautiful feather. The breed became very popular in the poultry farming industry within the British. He unfortunately never revealed how he interbred the chicken to create the modern-day Sebright, which is currently a rare breed. There are speculations that the breed is a mixture of the Hamburg, Polish, and Nankin chicken breed.

Safety

The safety of the chicken flock is paramount, and it could be threatened by many factors such as predators, climate, and illnesses.

The chicken coop should have a sturdy roof that conceals the chickens from the rain. Chickens do not enjoy the rain and are, therefore, important to protect them from it.

As we had earlier discussed, the elevated floor prevents predators from accessing the hens, but in the absence of that, a chicken farmer could create a stone layer beneath the chicken coop to block predators from accessing the inside of the chicken coop.

Ventilation

The chicken coop should at least have decent ventilation for the hens. Since there will be chicken droppings inside the coop, it would be important for the chicken coop to be aerated. This will also promote the health of the chickens and that the eggs produced are consumable.

Ventilation has also made it possible for one to place livestock in a confined space rather than placing them in a large space, thus taking up more land.

Ventilation ensures that there is fresh air coming into the coop to decrease the humidity and temperature to one that can be endured by the chickens.

The ventilation should be located above the chickens to protect the chickens from the wind ruffling their feathers and also protect the chicks from cold drafts of wind.

The ventilation of a chicken coop cannot be constant in a chicken coop, especially in a location where the weather is always changing. For example, the ventilation system for summer cannot be the same as that of winter. During summer, there will be excess heat, and the chickens will need to cool off while during winter, the chickens will need heat; otherwise, some may even die, especially the chicks.

It is recommended that there should be at least 2 ventilation points in the chicken coop for proper air circulation.

The Fundamentals of Ventilation

Scientifically, when chickens are confined in one area, there are different gases released, such as carbon monoxide, methane, ammonia, oxygen, and hydrogen sulfide. Some of the gases are from the decay of waste products from the chickens and their exhalation. These gases need to be replaced with fresh air, particularly oxygen, so that the chickens can breathe. The aeration is also healthy because it decreases the chances of infection of airborne diseases.

Ventilation decreases humidity within the chicken coop, thus making the chickens more comfortable. Heat causes air to expand, which gives it more volume to contain more moisture.

Humidity is a result of excess moisture in the air. Therefore, airflow within the chicken coop will ensure that the excess moisture exits the chicken coop. The availability of excess moisture would also be detrimental to the chicken coop, especially where the structure is not made out of waterproof material. This would negatively affect its durability over time.

Types of Ventilation Systems
There are 2 types of ventilation systems, which are the natural airflow system and the mechanical air movement system.

The weather is never constant, and as such, there is a need to apply both ventilation systems to ensure that the lives of the chickens are sustained no matter the weather.

• Natural Airflow System
This system relies on the airflow within the location where the chicken coop will be placed. Therefore, the placement must be strategic to maximize airflow, so the identification of the direction of the wind would be crucial in maximizing aeration within the chicken coop. Therefore, it would be advisable to place the chicken coop on higher ground so that the wind blows into the chicken coop. However, there could also be too much wind for the chicken, especially if it is cold, and as such, there would be a need for wind

barriers to regulating the amount of wind that can enter the chicken coop.

Winter ventilation systems are considered to be more technical than summer ventilation systems. This ventilation system would be crucial in a mostly cold area as it focuses on preserving heat within the chicken coop. In an attempt to preserve heat, there may also be a lot of harmful gases and humidity being confined in the chicken coop, which would be dangerous for the chickens. Therefore, there has to be a system in place that ensures such gases are released despite the cold. There have been several advancements in the winter ventilation system to the point that there are automatic systems in place that control the curtains and the windows to ensure that only the required amount of aeration is in place.

Summer ventilation heavily relies on the natural air system. The presence of side curtains on the chicken coop acts at the regulator on how much air may enter the chicken coop. The natural air gets into the chicken coop, thus eliminating the odor, excess heat, and moisture inside the chicken coop and replacing the gases with fresh oxygen for the chickens. However, ventilation faces challenges when the weather changes during the day, meaning that there is a need to combine the natural air system and the mechanical air system.

• Mechanical Ventilation System

This ventilation system is efficient because it ensures that there is aeration in the chicken coop no matter the weather. The system heavily relies on the electric fans to replace the air within the chicken coop.

There are 2 types, which are a negative pressure system and a positive pressure system.

Negative Pressure System

The fans solely release the air from inside the chicken coop, thus forming a vacuum within the chicken coop. The difference in pressure within the chicken coop and outside causes the air from the outside to be drawn into the partial vacuum to fill it. This, in turn, creates airflow within the coop.

Positive Pressure System

This ventilation system creates pressure within the chicken coop by the fans forcing air into the chicken coop. The difference of pressure outside and within the chicken coop cause airflow within to move through the outlets. Many chicken houses use a positive pressure system. There are 2 types of positive pressure systems. The first one thrusts the warm air into the chicken coop, which in turn pushes the air inside the chicken out. The second type causes the warm air to be shoved through plastic tubes with outlets into the chicken coop, thus mixing the air within.

The Mixture of Air through the Ventilation

A ventilation system should not only allow air to enter into the chicken coop but also ensure that the air mixes with the air inside the coop. This form of air circulation ensures that the air moves within the chicken coop to sustain life and also dilute other airborne diseases.

The absence of a proper circulation system means that the air inside the chicken coop will not have a proper circulation within which will lead to a fog within the chicken coop. Therefore, the fans should be strategically placed to allow for the movement of air within. The fans should be placed at the center of the structure, close to the ceiling, and on the furthest end of the chicken coop. This arrangement ensures that the warm air does not reach the ceiling because hot air rises, thus causing the air to move within the chicken coop.

There are also chicken coops that need additional heat, especially those located in warm areas. The requirements for this heat vary from one chicken house to the other. Some of the factors that contribute to this are the difference in the amount of aeration within the chicken coop, the temperature of the location of the coop, and the insulation standards.

The additional heat ensures that a farmer can regulate the amount of heat that the chickens need depending on the temperature outside the chicken coop. However, the mixture of air within the chicken coop ensures that the warm air moves within, thus generating heat to generate heat to warm the chickens during the cold season.

Elements of a Mechanical Ventilation System

• *Fans*

The fans push the air into the chicken coop, thus causing a mixture of air within the structure, which in turn causes airflow. Fans vary depending on the ventilation system needed for the chicken coop as well as the weather conditions of the area where the chicken coop is located. There is also the need for high-quality fans where the chicken coop is located in an area with very harsh weather so that it can effectively perform its function.

There is an association referred to as the Air Movement Association that has fan standards in place together with the means that the manufacture will use in establishing the performance rate of the fan they produce. It is, however, recommended to go for a larger fan than a small one because someone can at least regulate the fan if it is oversized, but a small fan would only be insufficient. That being the case, it would still be better to purchase the fan size that would be perfect for the chicken house owned.

There is also a guarantee that it issued where a fan is defective. The motors that run the fan need to be effective so that it is easier to manage. Therefore, the motors need to be connected in compliance with the available electrical codes of that region, which would mean that one needs a great electrician to do so.

• The Outlets and Inlets

This ensures that fresh air enters the chicken coop through the inlets to dilute the old air within the chicken coop, but due to this increases pressure from the fresh air, the old air is pushed outside through the chicken coop outlets. The design and placement of an inlet could fundamentally affect effective air circulation within the chicken coop. For sufficient air entry into the chicken coop, the velocity of the air ought to between six hundred to one thousand feet per minute. The slots should also be adjustable in width because the aeration requirement will depend on the chicken's weight, age, and temperature of the location. The adjustability is because these variables are never constant. The width of the slots should ideally be from 1 to 6 inches. The width spacing will be operated automatically or manually.

• The Control Setting

An example of a controlled setting would thermostats, which also work together with interval timers. This control setting is to ensure that there is a regulation procedure so that the temperature of the structure is conducive for the chickens.

One thermostat will have control over an individual single-speed fan. The control could also be over more than one-speed fan depending on the amount of ventilation needed. This thermostat will cause the speed fan to move where there is an increased temperature inside the chicken coop and deactivate it when the temperature decreases to a certain amount.

There are also different ventilation designs. Some have thermostats with duplicate switches that activate and deactivate fans that have 2-speed levels, thereby the speed that fan chooses will depend on the temperature of the chicken house. Some fans also have shutters that have a motor operation and should, therefore, be operated by the same thermostat so that these shutters work concurrently.

The interval timer is to ensure that there is no energy wasted, especially when the temperature outside the chicken coop is cold. The timer allows one fan to work while deactivating the other to ensure that the airflow within the chicken coop is sustainable for life but also warm enough for the chickens. Where the chicken coop is hot, the thermostat takes over the system to decrease the temperature inside.

The Material of the Chicken Coop

Chicken coop designs are, in most cases, made out of metal clay and wood. The material used in building the chicken coop will also depend on the needs of the farmer, and each of them has their benefits.

Chicken Coops Made Out of Wood

This material is sturdy for a chicken coop and is capable of being used to create a chicken coop with at least all the basic requirements. There are so many chicken coops that are made out of wood. However, it is important to note that the thickness of the wood will increase due to swelling during the rainy season. It is also a great material because it enhances air circulation within the chicken coop, thus making it warm enough for the hens.

However, a question does arise as to which type of wood would be best for a chicken coop.

The type of wood used in the construction of the chicken house will also determine its longevity because not all wood types will be durable. Some wood is not resistant to water and insects; however, there are a few ways that someone could curb this challenge. Below are a few of the ways to do so.

- Pressure can be exerted on the wood to inject insecticides into it to make sure it is insect resistant.

- Some woods do not decay, such as redwood, cedar, or tropical hardwoods, which would be ideal when constructing a chicken coop.

- There is also the option of treating softwood, such as hemlock, spruce, or pine. This means that the wood is infused with preservatives to ensure that it doesn't deteriorate.

- One could paint the plywood and use it on the exterior of the chicken coop.

Types of Wood

a. *Treated Lumber*

We shall specifically look into pressure-treated lumber. It is important to note that this wood has been infused with chemicals for purposes of preventing deterioration. Therefore, these same chemicals can end up being infused into the chickens, which can later enter into the by-products of the chickens. This means that a human being could consume these same chemicals unknowingly, which would be dangerous to their health over time.

It is, however, crucial to note that when choosing lumber, there are different grades. The difference in the grade of lumber will establish whether it shall make a sturdy structure. Where the grade is labeled to be of a smaller number or "standard and better" it means that the wood will be strong, making a sturdy chicken coop. There is also lumber that is referred to as unseasoned, which absorbed a lot of water, thereby making it difficult to be induced with preservatives or paint application. Kiln-dried lumber is dry wood, which means that it doesn't easily absorb moisture and will, in turn, bond with preservatives better upon application.

That being said, we should first begin by looking into the demerits of pressure-treated lumber which are as follows:

- Pressure treatment can lead to the infusion of copper into the soil. This would be harmful, especially where the soil is being used to grow crops that human beings or other animals could benefit the consumer. However, the chemicals being infused onto the wood have improved over time and are more health-conscious. Despite these improvements, the United States Environmental Protection Agency guides that chicken coops that have been made out of pressure-treated wood should not be placed on the soil, where crops are growing or in a place where other animals could access from the exterior.

- There is an additional cost when purchasing locks or screws because one cannot use regular screws on pressure-treated wood. This is because chromate copper arsenate, which is the chemical used in treatment, will cause the fasteners to rust, thus needing constant replacement and weakening the structure. Therefore, one would need to use hot-dipped galvanized fasteners or stainless steel to ensure they do not corrode, and they are more expensive than regular fasteners.

The advantages of using pressure-treated lumber are as follows:

- The best characteristic of pressure-treated wood is that the preservatives used on the wood ensure that it is more durable. This is because it is insect resistant, it doesn't deteriorate easily, and it is waterproof. This ensures that the chicken coop can endure harsh rains, and a farmer can comfortably clean their chicken coop with water without fear that it will deteriorate over time.

- Pressure-treated lumber is much cheaper in comparison to other types of wood.

- There is the challenge that the preservatives could leach into the soil, which would be a health hazard where crops are being grown for consumption. However, one could curb this challenge by painting the wood.

b. *Tropical Hardwoods, Cedar and Redwood*

Other than going for the man-made durable wood, one could opt to go for a chicken coop that is made out of naturally durable wood, which would be the three mention above. It would also be important to note that redwood and tropical woods are not available in certain places as compared to cedarwood, and as such, in comparison to the 2, cedar is used more when making chicken coops.

The following are the advantaged or tropical hardwoods, cedar, and redwood.

- These types of wood do not corrode; therefore, they will naturally sturdy structures that will last for a very long time.

- Since they are not pressure-treated woods, they are healthier for the chickens and the environment.

- One would not have to paint the wood consistently so are to prevent chemicals from leaching into the soil, making it less time-consuming.

The disadvantages of this chicken coop are as follows:

- These types of wood do sound like the better option, and as such, there is a high demand for them, thereby making them very expensive. This makes a chicken coop made out of these types of wood even more expensive.

- Cedar is popular for its non- decay characteristics; however, this doesn't apply to all forms of cedar. There are heartwood and sapwood cedar. The old heartwood cedar is the one that has the rot-resistant characteristic, but it is not easily available in comparison to heartwood cedar. Therefore, if one cannot acquire the heartwood cedar, they would be forced to treat it. That being the case, most people would prefer to purchase softwood where they cannot purchase heartwood cedar because it would be cheaper.

c. Preserved and Painted Softwood

This is the recommended form of wood to use in comparison to the ones mentioned above. The reason for this is that it is cheaper and there is a way of avoiding the toxins that would be emitted from pressure-treated lumber.

Normally, softwood, such as spruce or hemlock, are treated with paint that doesn't have any toxins in it, that could leach into the soil or be absorbed into the chickens. Their wood is also treated with preservatives to ensure that it doesn't deteriorate easily. This means that in comparison to the pressure-treated lumber, the preserved, sealed, or painted softwood is the better option because it is not toxic. However, one could also attempt to ensure that pressure-treated lumber is created using non- toxic chemicals instead. This wood is even better than the tropical hardwoods, cedar, and redwood because it is a cheaper option.

The advantages of this wood are as follows:

- There are different chemicals used when preserving softwood, but the recommended would be TimberPro UV's Internal Wood Stabilizer because it increases the density of the wood, thus making it sturdier. This liquid preservative is also easy to apply, and one only needs 2 to 3 layers of paint. The preservative does make the wood undergo some color changes over time because it starts as a vibrant color and dwindles to a silver color. There is a case example of a person who stated that their coop made out softwood had been around for close to a decade, and they reside in a rainy place which is Portland, Oregon. Further, this area has a very hot season, and the coop has been able to withstand it after all these years.

- There will be no extra expense to purchase hot-dipped galvanized fasteners or ones made out of stainless steel because the basic fasteners would nor corrode in the wood as compared to the pressure-treated lumber.

- One has the option of painting over the wood is for decorative purposes. However, since the wood has been treated, the paint should be given enough time to dry for approximately 3 weeks. This is because the Internal Wood Stabilizer protects the wood well enough from any paint used on top of it. The part of the chicken coop that will be exposed to water is the one that should be protected using the Internal Wood Stabilizer, such as the exterior of the chicken coop, which would withstand the rain. It is, however, important to note that most farmers tend to use paint rather than apply the Internal Wood Stabilizer application, which will seal the wood from heavy rainfall and sunlight, yet there would be fasteners used on the wood. These fasteners will expose the wood, thus allowing moisture to penetrate the wood leading to it rotting quickly and, as such, negatively affecting its durability. Therefore, avoiding applying the Internal Wood Stabilizer will be detrimental to the integrity of the structure. The paint used on the chicken coop should also be non-toxic to the animals and the soil, especially because human beings consume the by-product of these two resources as a source of food.

The demerits of preserved and painted softwood are as follows:

- The application of the preservatives is time-consuming. However, a treatment such as TimberPro UV's Internal Wood Stabilizer shouldn't take too long since it is in liquid form. There is also time wasted where one opts to paint over the wood because they would have to wait for approximately 3 weeks.

- The Internal Wood Stabilizer is not an insect repellant. This, in turn, will affect the durability of the structure because insects such as termites will consume the wood, thus weakening the chicken coop over time. This will lead to an extra cost of replacing the entire structure.

d. Plywood

There are certain areas of the chicken coop that may require plywood to be used. However, plywood comes in different forms with unique characteristics, which will also mean that they will also come at a different financial cost.

The following are the different types of plywood:

- Oriented Strand Board

This wood is created by using an adherent and pressing sheets of wood strands to form a uniform layer. This plywood is also the cheapest one available. Its cheapness can be attributed to the fact that it is not water-resistant and, as such, would easily deteriorate in the presence of moisture and need replacement. Therefore, this plywood would be best used on the interior of the chicken coop, and where it is placed at the base of the chicken coop, it should be sealed with linoleum.

It is also important to note that the basic plywood is not preferable to use on the outside of the chicken coop because it can only withstand minimal amounts of moisture, which would come from the weather. This is because the glue on the plywood used can only withstand minimal moisture. Therefore, where a farmer intends to use standard plywood on the exterior of the chicken coop, they ought to ensure that it is sealed with a primer and 2 coats of latex paint on the exterior. Hence, plywood would be more useful to be used in the interior of the chicken coop.

- Marine-Grade Plywood

To begin with, this plywood is not easily available in the market as compared to regular plywood, and it is also far much more expensive. The expense can be attributed to the fact that the adherent used is impermeable to water, thus making it a long-lasting structure since it will not swell and break down. This plywood would also make a long-lasting structure.

- Medium Density Overlay Panels

This plywood is the middle ground between the marine grade plywood and the oriented strand board. It bears the benefits of the marine-grade plywood, and it is also cheaper. This plywood is covered with resin, which is pressurized onto the wood using heat to bind it to the plywood. The resin sealant used makes the plywood water-proof, thus preventing the wood from rotting over time, thus ensuring durability. The added advantage is that the resin overlay gives the plywood a smooth surface, thus making it easy to paint on for aesthetic purposes.

Chicken Coops Made Out of Metal

Metal is sturdier than wood and plastic; however, it is recommended that farmers should strive to avoid chicken coops that are made out of metal. The reason or this is that metal absorbs a lot of heat, which would not be conducive for the hens. There is also the possibility of metal rusting due to the water from cleaning and the rain. Metallic chicken coops also do not have proper air circulation. Despite all these cons, a corrugated steel roof is extremely effective on the chicken coops.

Chicken Coops Made Out of Plastic

Plastic chicken coops weren't as popular, but over time they seem to be increasing in the market. The chicken coop will perform its function but for limited time because plastic chicken coops are not long-lasting in comparison to those made out of plastic or wood. The chicken coop will have ventilation, but not as much as a wooden chicken coop. During the hot weather, the plastic chicken coop will have better insulation than the metallic chicken coop. The best feature of a metallic chicken coop is the hygiene aspect because it is easier to disinfect, and there will be no rusting. Plastic doesn't do well during the cold weather, but it conducive during other weather conditions.

Chapter Four

The Rationale behind Purchasing Chicken Coops

Numerous companies have come up with different chicken coop designs that can meet the needs of the intending purchaser. The features of the chicken coop will depend on the price tag attached to them. This, however, doesn't mean that a fully loaded chicken coop design will be unreasonably expensive. However, an intending purchaser has the option of building a chicken coop but several reasons, but purchasing a chicken coop would better option for most people. Below are a few of those reasons.

- Purchasing a chicken coop is less time consuming.

Building a chicken coop takes a lot of time, depending on the design hat the farmer wants. There are several steps that a farmer would have to undergo, such as drawing the design, looking for the raw materials, modifying the materials such as wood, amongst others. These steps take up a lot of time; for example, one could have softwood but would have to find a way to make sure it is treated to ensure that it will make a durable structure. There is also a lot of time spent in research to find out the best materials to use on the chicken coop and what it should entail. There are also pieces of equipment within the chicken coop that the builder has to construct as well, such as the nesting box, the perches, and the run. However, these chicken coop requirements can be readily available to a person who intends to own a chicken coop; all they simply have to purchase is one. It would take less time and energy to purchase a chicken coop rather than building one.

- Purchasing a chicken coop will give you more relaxation time.

Most people have so many duties that require their attention, such as their job, family, and other activities; therefore, they barely have time for themselves. Some people would find pleasure and relaxation in building something from scratch, but for other people, it would be more stressful. Therefore, the idea of having to build a chicken coop might even discourage them from owning chickens. For such people, it would be easier to purchase a chicken coop, which will still have all the features of a built chicken coop depending on its design.

- Building a chicken coop can be expensive.

When purchasing a chicken coop, people can compare prices, buying one immediately or budgeting for one in the future. When building a chicken coop, there no specific amount in place. Therefore, building the chicken coop can end up being more expensive than purchasing one. The only way that a builder would know the cost of building the chicken coop is if they have built one before or if someone who has previously done it helps them. There are also certain material and tools that are very expensive to purchase such as hardwood or an electric gun among others

- Purchasing a chicken coop is easier to assemble than building one.

Most chicken coop designs in the market come with an instruction manual on the means of assembling it. These instructions do not need a certain amount of skill to follow, meaning that anyone can do it. Some chicken coops even take minutes to assemble, while others take a few hours, but building one could take days, weeks, and even months depending on the design and the material used.

- Whether one buys or builds a fully loaded chicken coop, the chickens will still be equally comfortable.

The chicken coop design available in the market varies; however, many designs come fully loaded. This purchased chicken coop will have all the needs to sustain a happy chicken, especially in the chicken coop design. It is fully loaded with nesting boxes, perches, an attached run, a proper ventilation system, among others. Therefore, if one purchases a chicken coop, they will spend less time and energy and achieve the same result in terms of the happiness of the chickens as someone else who chooses to build one.

- Purchasing the chicken coop will allow the owner to know the features of the chicken coop beforehand.

Many physical and online stores have great customer care services to inform the purchaser on the features of the chicken coop and the benefits of having them in a specific design. For example, a customer care agent will advise a customer on the importance of having removable trays in the chicken coop, which is to collect chicken dropping. The customer care agent could also inquire into the climatic conditions of the location where the chickens will be to establish the perfect ventilation system for the chicken coop. However, when building a chicken coop, one may not know the new features available that will ensure the safety and comfortability of the chickens.

- A purchased chicken coop has been used by numerous people, and therefore one would establish whether it is of good quality or not.

Most online pages that sell chicken coops have review sections for their product. This will allow a purchaser to know whether the chicken coop lives up to its name. Another way of acquiring such knowledge is where someone else who has previously used that chicken coop designs informs you whether it was worth the money spent. The added advantage is that most chicken coop designs come with a warranty; therefore, one could return it if it is defective. Building your chicken coop means that one may not know the challenges that the design beforehand. The built chicken coop might also be defective, thus forcing a farmer to spend their own money to fix it.

- There is a guarantee that the chicken coop features will perform their intended functions.

The chicken coops available in the market have already been tested to have certain capabilities. For example, Pawhut wooden customized chicken, a chick coop, is made out of fir wood and has been proven to be water and insect resistant, thereby having a durable characteristic. Therefore, a purchaser will not have doubts about the performance of the chicken coop.

- Purchasing allows a customer to see whether the chicken coop is appealing or not.

There are very many stylish and beautiful chicken coop designs in the market that an intending chicken owner can purchase. This also allows the purchaser to choose a chicken coop that would fit the aesthetic of their backyard and house. There is also the concern that the built chicken might be poorly constructed, thus looking terrible in the backyard.

Chapter Five

Top 50 Chicken Coop Designs for 2020

In the recent article The 6 Best Chicken Coops of 2020 (the Spruce, 27th April 2020) it listed the top 6 chicken coop designs. Those shall be the first chicken coop designs that we'll look into. The article further noted the key features of each design that stood out for it to make it on that list.

The article ranked the best one of them all to be the *SnapLock Fromex Large Chicken Coop*, which can be obtained on Amazon. The rationale behind this was due to its usability, safety to the chickens and that it is weatherproof.

The article noted the chicken coop that would be best for beginners was the *Whatever Works Chicken Coop with Planter* at *Whatever Works* because of its flexibility for expansion of the poultry in the future.

The most conducive chicken coop for urban areas was the *Omlet Eglu Chicken Coop* at *Omelet*.

The most expensive was the *Williams Sonoma Cedar Chicken Coop & Run with Planter* at *Williams Sonoma* with 2 feet running space and made cedar, which doesn't rot. However, it can only accommodate 6 chickens

The best chicken coop with multiple levels was the *PawHut Outdoor Wooden Chicken Coop with Nesting Box and Run* at *Walmart*, which has a high nesting box and a ramp.

The article concludes with the best chicken coop to walk into is the *Round Top Stand Up Chicken Coop* at *Roost and Root*. It has a door, and it is high enough for a farmer to move around when inside.

Therefore, this paper shall further look into these top 6 chicken coop designs before expounding further into other designs.

1. SnapLock Formex Large Chicken Coop

This chicken coop can house 6 chickens, but this solely depends on the type of breed. The chicken coop is obtained from Amazon.com. It is also more expensive than some wooden chicken coops, but it comes with incredible usability and assembly. The assembling of the chicken coop could even take up to 15 minutes.

The chicken coop comes with unique features such as:

- Two Nesting boxes have been split into 2, thus providing 4 nesting spaces for the chickens. The nesting boxes are further removable, thus making it easier to access the eggs and clean the nesting area.

- It is hygienic due to the availability of a removable tray to collect the chicken droppings.

- It is water-resistant and has UV light protection, which contributes to its long-lasting characteristics long-lasting.

- It is also convenient because it is made from dc double-wall plastic, which is can also be used for storage.

- Proper ventilation due to the adjustable side vents.

- It is a strong structure and thus can be illustrated by a farmer who reported that a bear attempted to eat the chickens but couldn't get access to them because of the well-built structure.

The first demerit with this chicken coop is that it doesn't have a run, and as such, the farmer must create or purchase one so that the chickens can move around, thus increasing their comfort.

2. Whatever Works Chicken Coop with Planter

This chicken coop is great for a beginning poultry farmer. The chicken coop is flexible as it offers more room for a farmer to increase their flock over time.

The chicken coop is also conducive for different weather conditions because it is made out of wood and has coated roof panels. This guards the chickens against the rain, because chickens do not like the rain, and extremely cold weather could result in the death of the chicks.

There is also ample air circulation within the chicken coop and the presence of sunlight for the chickens as a source of vitamin D, which also contributes to the egg-laying process.

A unique feature that it has is the availability of a planting section where a one could grow a few plants.

This chicken coop is also referred to as "poultry palace" and can hold approximately 3 to 5 chickens.

A distinct feature of this chicken coop is that it has multiple levels, thus providing ample space for the chickens to move around. This particular chicken coop is 69.25 x 26 x 43.5 inches, thus providing enough roaming space.

The chicken coop has several features including:

- An elevated nesting box for the chickens to comfortably lay their eggs.

- A non-slip ramp for the chickens to access the living space.

- A run which can be accessed from the side door for the chickens to roam.

- A hinged roof which is located above the nesting box so that a farmer can easily have access to the eggs.

- It has a tray that slides out to collect the chicken droppings and, as such, promotes cleanliness within the chicken coop.

This chicken coop can be bought from WalmartBuy on Aosom.com.

3. Omlet Eglu Chicken Coop

The appearance of this chicken coop is quite stylish and, as such, suited for the urban poultry farmer. Therefore, it would be a great backyard chicken coop. It also has amazing usability, so the farmer wouldn't have much difficulty in maintaining the chicken coop.

The features Omlet Eglu Chicken coop are as follows:

- It is a portable chicken coop, and it isn't heavy for a farmer to move it around.

- It is made out of plastic and is green in color.

- It has a run made out of steel mesh, which has the chicken and water feed containers attached to it.

- There are also doors, panels, and a drawer so that the poultry farmer can easily clean it since it is accessible. Another benefit of this is that it provides ventilation as well within the chicken coop due to these openings.

The demerit with this chicken coop is that it is expensive in comparison to others, but its stylish design makes it an accessory in your backyard.

This chicken coop is available on Omlet.us.

4. Williams Sonoma Cedar Chicken Coop & Run with Planter

This chicken coop is also very appealing in the eye and can house 4 to 6 chickens depending on the breed acquired and, most importantly, if the poultry farmer wishes to allow them to move around during the day.

The characteristics of this chicken coop are as follows:

- It is made out of cedarwood, making it a long-lasting structure, thereby eliminating the concern of having to replace the structure.

- It provided safety to the hens by protecting them from predators due to its wire mesh and galvanized roof.

- It is a multi-layered chicken coop.

- The size of this chicken coop is 63.25 x 61.75 x 83.25 inches.

- It has waterproof nesting boxes, which are where the chickens lay their eggs.

- It also has a run for the chickens to roam around, which measures up to 25 square feet.

- There is also an additional planter for someone to grow whatever they wish.

This chicken coop is available on Williams- Sonoma, which also has a great delivery and assembly service.

This chicken coop is available on Williams- Sonoma.

5. Round Top Stand Up Chicken Coop

The size of this chicken coop is 58 x 66 x 85-inches; therefore, it has enough space for the flock.

The other features of this chicken coop are as follows:

- The distinct feature about this chicken coop is that it is 85 inches in height, thus making it easier for a person to access the chicken coop without having to bend over.

- The chicken coop can accommodate 2 to 6 chickens.

- It has nesting boxes for the chickens to lay their eggs. The availability of nesting boxes also makes it easier for the farmer to collect the eggs.

- It is also flexible due to its run, which can be used for future development since it is on the outside.

Additional Chicken Coop Designs

6. Pets Imperial Double Savoy Chicken Coop

This chicken coop can hold between 6 to 10 chicken, all depending on the breed being reared.

It provides the farmer with all the basic requirements that every chicken coop ought to have with the following features:

- It has 2 nesting boxes so that the chickens feel secure enough to lay their eggs and ease for the farmer in collecting them. The chickens will also feel comfortable doing so because the 2 nesting boxes are divided into 6 sections; therefore, there is enough resting room for the chickens.

- There is ample space for the chickens, thus increasing their comfort.

- There is a roof that can be opened, which will help during the cleaning of the chicken coop.

- There is also a tray for collecting the chicken droppings, thus increasing the hygiene standards within the chicken coop. The tray is made out of galvanized metal, which will not erode quickly.

- It has 4 perches for the chickens to rest on, and at the same time, it protects them from underground predators.

- It is durable, as it can last very many years.

- It is also very easy for a farmer to assemble.

- It has an elevation to protect the chickens from underground predatorial attacks.

- There is great aeration within the chicken coop.

- The chicken coop is made out of treated wood, thus making it sturdy and can withstand harsh weather, and this particular wood is protected from insect destruction.

The challenge with this chicken coop is that there is no run for the chickens to move around in. This forces farmers to build or purchase a separate run for the chickens. It is, therefore, important to note that the manufacturers of this coop do have separate runs available for a farmer to purchase.

7. Petsfit Weatherproof Outdoor Chicken Coop

This chicken coop has a lovely appearance that resembles a barn. The best feature of this chicken coop is its weather-resistant capabilities. That being said there are other great features that it has such as the following:

- It has nesting boxes for the chickens to rest and lay their eggs comfortably, but it only has 2 compartments.

- There is ample space; thus, there will be no overcrowding, which would be a health hazard to the chickens.

- The chicken coop is made out of cedarwood, thus making it more durable and protects the chicken from the rain. The paint used is also water-resistant, thus contributing to its durability.

- There is ample ventilation that provides the form of openings.

- The floor is removable, thus increasing the hygiene standards because it is easier to clean.

- It also has an endothermic roof.

- The chicken coop is also easy to assemble.

The demerits of this chicken coop are as follows:

- The roof cannot be opened, and as such, there is limited accessibility for the farmer during cleaning. Typically, a hose is used for cleaning and the water to drop to the bottom instead.

- The chicken coop can only accommodate three to four chickens.

- The company that issues this chicken coop only issues 30 days of warranty, after which the farmer will bear the cost if it is faulty, which is a very limited time to detect defects.

- Over time, the paint deteriorates, thus forcing the farmer to have to repaint the structure.

- It does not have a run; therefore, a person would need to build or purchase a separate run for the chickens to roam around.

8. BestPet Chicken Coop

Where a poultry farmer already has chicken coop, this chicken coop would be a great addition to an already existing chicken coop.

This chicken coop has the following features:

- It is spacious for the chickens; therefore, there isn't overcrowding inside the chicken coop.

- There is also a run for the chickens to roam. This chicken coop would be a great addition, especially if the already existing chicken coop does not have a run.

- The chicken coop is also easy to assemble.

- This chicken coop is durable because it has an amazing coat finishing that protects it from deteriorating and rusting. It is also long-lasting because the frames are made out of steel, which gives it UV light protection, thus reducing the chances of breaking down.

- The chicken coop is also multi-purpose because rabbits and ducks can also use it.

- The coop also protects the hens from harsh weather such as extreme heat, rain, and snow.

- It is a portable chicken coop and can, therefore, be moved around at the convenience of the poultry farmer upon being set up.

- The downside of this chicken coop is as follows:

- It cannot accommodate many chickens; therefore, it is recommended to have more than one.

- The chicken coop is better as an additional chicken coop to an already existing one.

- This chicken coop does not have nesting space, which is crucial for the chickens to rest and lay eggs. The other challenge drawn from this is that it will not be easy for the farmer to collect the eggs in the absence of nesting boxes.

9. Best Choice Products Outdoor Wooden Chicken Coop

This would be a great chicken coop for those who are beginners because it has all the necessary features to begin with.

The features are as follows:

- The chicken can withstand harsh weather conditions such as rain, which they do not like because it is made out of treated wood and galvanized chicken wire. This, in turn, makes it a long-lasting structure.

- It would also be great for a beginner because it can only accommodate 3 to 4 chickens.

- The chicken coop comes with a nesting box for the hens to lay their eggs and sleep.

- The roof can be opened, thus making it easier for the farmer to clean, collect the eggs, and feed the hens from above.

- There is a small space outside for the hens to roam around.

- It is multipurpose because it can also be used to accommodate other animals such as ducks and rabbits as well.

- It has an elevation to protect the hens from underground predators.

- The chicken coop is safe for the chickens because it has 2 doors which one can lock, making it also accessible due to duplicate entry points.

- It has a tray that can collect the chicken drooping, which can be slid by the farmer. This, in turn, improves the hygiene standards within the chicken coop.

- There is ample ventilation because it has a metallic wire mesh, thus allowing air circulation and, in turn, prevents the chickens from suffocating.

- There is a ramp for the chickens to move on to access their nesting spaces.

10. Best Choice Chicken Coop

The best thing about this chicken coop is that it has the basics that a poultry farmer would need when raising chickens.

The features of this chicken coop are as follows:

- A nesting box has 2 compartments for the chickens to rest and lay their eggs. The farmer would also have ease when collecting the eggs.

- It would be a great chicken coop for beginners because it can only accommodate two to 4 hens, and it is also easy to assemble.

The challenges with this chicken coop are as follows:

- The absence of a chicken run for the hens to be roaming around, thus forcing a farmer to either purchase or build one.

- There is the possibility of theft of the eggs by other animals and even attach from predators such as snakes because the door doesn't have a locking system place, Therefore, if a farmer is in an environment that has animals around, the chickens would not be safe inside this coop. However, this challenge can be curbed by installing a locking system on the hinged door.

- There would be an extra cost to treat the wood to protect it from harsh weather conditions.

11. Advantek-The-Stilt-House

This is a chicken coop that has great quality. The features that this chicken coop possesses are as follows:

- The availability of a run outside for the chickens to move around, thus contributing to their overall happiness and can be used by other animals other than chickens as well.

- The chicken coop is multi-purpose, meaning that one to house other poultry other than chickens such as ducks.

- Feeding the chickens is easy on a farmer of the double split nesting case for food and water.

- There is a tray that a farmer can pull to collect the chicken drooping, thus increasing hygiene within the chicken coop.

- The chicken coop is made out of fir wood that is protected from insects and deterioration.

- The chicken coop has a reddish-brown color, which contributes to its stylish appearance.

This product can be obtained from Amazon.com.

12. Tangkula-Chicken-Wooden-Garden-Backyard

This chicken coop was created by Tangkula and has come to be one of the best chicken coops in the year 2020.

The features that this chicken coop possesses are as follows:

- The chicken coop can house numerous chickens in comparison to others.

- The chicken coop is appealing to the eye with a green shade roof and wooden color.

- The room for the chickens to rest is warm, which is most important, as it is healthier for the chicks because they could die in the cold.

- There is a slip tray for collecting chicken drooping, thus promoting the hygienic standards inside the chicken coop.

- The chicken coop is spacious, thus ensuring there isn't overcrowding, which is safer for the chickens and also ensures their comfort.

- The structure is durable because it is made out of cedar, which is rot resistant.

The disadvantage is that cedarwood is substantial.

This product can be obtained from Amazon.

13. Chicken Coop Outlet-Large-Wood-Chicken-Coop

This chicken coop can accommodate at most 6 chickens depending on the size and breed on the hens.

This chicken coop has the following features:

- The coop is great for the outdoors because it is made out of spruce wood, which reduces the possibility of the wood chipping, thus ensuring that it is long-lasting time.

- There is a tray made out of zinc for the collection of chicken droppings, thus increasing the hygiene of the chicken coop and makes it easier to clean.

- The structure has sufficient ventilation to ensure that the chickens do not suffocate

- The delivery of the structure by the company is done on time.

The disadvantage of this chicken coop is that it can only be shipped within the United States.

This product is available on Amazon.

14. PawHut-Wooden-Portable-Backyard-Chicken

This is one of the best chicken coops available in the market.

The advantages of this chicken coop are as follows:

- There is an outdoor run for the chickens to roam around and contribute to their overall happiness.

- The nesting box has been well distributed for the chickens to lay their eggs comfortably.

- The structure is made out of spruce wood, which protects it from chipping, thus ensuring that the structure is long-lasting.

- The structure has a galvanized chicken wire mesh and locks on the door, thus ensuring the safety of the chickens.

- The structure can withstand different climatic conditions.

- There is a ramp for the chickens to access their outdoor run and nesting spaces.

- There is also sufficient air circulation within the chicken coop due to the availability of side slots.

- There are also perches for the chickens to rest one because, in the wild, chickens would naturally rest on perches to protect themselves from predators.

- The product is always stocked by the manufacturing company; thus, it is always available to interested buyers.

- The chicken coop has wheels, thus making it portable, thereby meaning that a farmer could move the chickens around, making them comfortable, which in turn leads to a better product.

The disadvantage, however, is that it can only house 3 chickens.

This product is available on Amazon.

15. Merax-79-Inch-Animals-Nesting-Chicken

This chicken coop has the following features and advantages:

- The product has high quality and affordable to any interested purchaser.

- The is ample air circulation due to the availability of a window.

- There is a large nesting box for the chickens to lay their eggs and sleep.

- There are 2 rods the structure for the chickens to rest on, which makes them much happier and protects them from underground animals. The rods would make it feel more natural to them.

- There is a tray that collects the chicken droppings, which is removable, thus making it easier to clean.

- The structure is also very stylish.

The downside of this chicken coop is that it is not available to every country and that the wood can easily get damaged.

This product is available on Amazon.

16. Certainty-Rabbit-Hutch-Chicken-Coop

One distinct feature of this chicken coop is that it provides ample space for the chickens to roam, thus contributing to their comfort.

This chicken coop has the following features:

- There is ventilation for the chickens, thus reducing the chances of suffocation.

- The chicken coop is made out of wood, thus benefiting from the advantages of a wooden chicken coop, as we had earlier discussed, such as being weather-resistant and sturdy.

- The chicken coop sides are adjustable.

- The structure is also beautiful in design.

- The chickens can easily move around.

The disadvantage is that the company that makes the product does not issue an instruction manual to the purchaser on how to assemble the chicken coop.

17. Pawhut-122″ Best-PawHut-Chicken-Coops

This chicken coop would be considered one of the most suitable chicken coops in the market.

The features of this chicken coop are as follows:

- The structure is long-lasting because it is made out of fir wood, which protects it from deteriorating and insect resistant. This material also protects it from harsh climatic conditions.

- The chicken is flexible, and one could, therefore, accommodate other types of poultry.

- The structure is secured by a wire that borders the first level of the chicken coop, thus providing security from predators.

- The chicken coop is also well ventilated.

The advantages of this chicken coop are as follows:

- It is portable, meaning that it can be moved around.

- The strong cables provide security to the chickens from predators.

The only disadvantage is that the instruction manual provided is not well written for purchase to assemble upon buying the product easily.

This product is available on Amazon.

18. White Coops and Feathers Extreme

This chicken coop is larger and can comfortably accommodate 8 hens. This model has some unique features, but it is expensive. The features of this chicken coop are as follows:

- The chicken coop is spacious for the chickens to roam.

- The chicken coop has multi-layered contributing to the spacious features.

- The ground level has a shade that protects the chickens from UV light and ensures that it is not too hot for them, thus making them comfortable.

- The second floor of the chicken coop provides the chickens with their sleeping and nesting space.

- A farmer can access the nesting space from the outside, making it easier to collect the eggs.

- There is also a metallic tray that is removable for the collection of chicken droppings, thus promoting the hygienic standards of the chicken coop.

- There is ample ventilation which can be closed where it is needed, for example during very cold weather.

19. Yardeen Coop

This chicken coop has the following features:

- This chicken coop is made out of wood, thus benefiting several features such as being impermeable to water and a durable structure.

- The roof further has asphalt shingles which contribute to its weather-resistant feature. This will ensure that the chickens are protected from the rain, which they are not fond of.

- The chicken coop can accommodate 6 chicken.

- The chicken coop is adaptable in the sense that it can accommodate other animals such as rabbits or ducks.

This product is available on Amazon.

20. Trixie Chicken Coop

Trixie Chicken Coop can accommodate up to 6 chickens, but it can house 10 small chicken breeds such as bantam chickens. Other than that, it also has other additional features which are as follows:

- The chicken coop has an outdoor run for the chickens to roam around.

- There is also a door that ensures that the chicken coop is secure by protecting from predators.

- There are 2 doors, the upper and anterior door, which makes it easier for the farmer to access the chickens.

- The upper door has 2 resting houses for the hens.

- There are also removable perches for the chickens to comfortably move around whenever the farmer wants to create more room for the chickens.

- There are also ramps for chickens to move back and forth in the multi-layered chicken coop.

21. Confidence 62-Inch Chicken Coop

One pronounced feature of this chicken coop is that it is spacious for the chickens to move around and be comfortable.

The features of this chicken coop are as follows:

- There is ample air circulation, thus ensuring that the chicken coop is not stuffy.

- The chicken coop is made out of wood, thus making it long-lasting.

- The chicken coop also has a stylish design.

- The chicken coop is small and will not use a lot of the land available.

- The chicken coop is also easy to maintain proper hygienic standards.

This product is available on Amazon.

22. Lazy Buddy

Lazy Buddy is spacious, thus ensures that there is no overcrowding amongst the chickens, which makes them more comfortable and protects them from harm amongst each other.

This chicken coop has the following features:

- The chicken coop is elevated, thus protecting the chickens from underground predators.

- There are also nesting boxes for the chickens to comfortably lay their eggs and the farmer to collect them easily.

- The chicken coop is made out of wood, thus making the structure firm and water-resistant as well. This material also makes the structure long-lasting and ensures that the structure does not deteriorate.

- The paint used on the structure is not harmful to the chickens.

- The chickens can easily move around because the chicken coop comes with 3 doors.

- The structure is easy to clean because it has multiple opening that ease of access during cleaning.

- The structure is easy to set up and will only take approximately 20 minutes of your time.

This product is available on Amazon.

23. Pets Imperial Monmouth Large Chicken Coop

This chicken coop has undergone numerous renovations before achieving its current status today to deliver a better product to the purchasers.

This present design has the following features:

- This chicken coop has a removable galvanized metallic tray for the collection of chicken dropping, thus making it easier to clean.

- There is also a surface underneath the metallic tray to support the chickens when the coop is being cleaned.

- The chicken coop can accommodate 3 to 4 chickens.

- The chicken coop is made out of fir timber, meaning that the structure is durable because it cannot be damaged by insects and doesn't rot.

- There is also a galvanized mesh to protect the chickens from predators such as foxes.

- There is a roof that can be removed, thus making it easier to clean the chicken coop.

- There is a nesting box making it easier to collect eggs.

This product is available on Amazon.

24. New Age Eco-Flex Fontana Chicken Barn

This barn-inspired chicken coop is one environment-friendly chicken coop because it is made out of reused polymers and wood by-products.

The structure has the following features:

- The structure is durable because it is made out of Eco-Flex, which doesn't have the expansion and contraction characteristics of wood, thus making it a sturdy structure. The material also makes it water-resistant, and there will be no cracks in the structure over time

- The chicken coop being water-resistant makes it easier to clean because it will not absorb water.

- The roof of the chicken coop can be removed to collect the eggs and cleaning of the chicken coop easily.

- There are gaps available for the circulation of air within the chicken coop.

This chicken coop is available on Amazon.

25. PawHut 88" Wooden Backyard Slant Roof Hen House Chicken Coop

PawHut 88" wooden chicken coop meets all the basic requirements that any chicken coop ought to have.

The features of this chicken coop are as follows:

- An outdoor run for the chickens to roam around comfortably. The run is further fences to protect the chickens from predators.

- The chicken coop is spacious for the chickens to live in comfortably.

- The is controlled ventilation, meaning that air can be limited, especially during the cold weather.

- The nesting box has 2 divisions for the chickens to lay their eggs comfortably and for the farmer to collect them easily.

- The nesting box for chickens to comfortably lay eggs.

- There is easy accessibility for cleaning purposes because there are multiple entry points

- There is a tray for the collection of the chicken dropping, which can be removed, thus promoting the cleanliness of the chicken coop.

This product is available on Amazon.

26. Eco Liner Outdoor 80" Wooden Chicken Coop

This chicken coop is attractive but also serves its intended purpose, which is chicken rearing.

This chicken coop has the following features:

- The chicken coop is spacious to accommodate the chickens comfortably.

- The nesting boxes are large enough to accommodate 2 chickens. The availability of a nesting box means that the chickens can comfortably lay their eggs, and the farmer can easily collect them.

- The structure has a window to promote ventilation within the chicken coop.

- There is a removable tray for the collection of the chicken droppings ensuring that it is easy to clean.

- The chicken coop is made out of wood, thus attracting the best features of this material, is that it is durable.

- The roof is made out of asphalt to protect the chickens from the rain.

- There is also a spacious run for the chickens to move around.

- The chicken coop is easy to set up.

- The chicken coop is surrounded by a galvanized wire mesh which secures the chickens by protecting them from predators.

- The chicken coop can accommodate 2 to 4 chickens depending on the size of the breed.

This product is available on Amazon.

27. PawHut 114" Wooden Customizable Backyard Chicken Coop

The PawHut 114" is a great chicken coop that would ensure the chickens are happy and comfortable.

This chicken coop has the following features:

- This chicken coop is adaptable because it has a removable duplicate exterior that one can redesign to their personal needs.

- It has is a multi-tier chicken coop, thus having plenty of room for the chicken.

- There is a ramp for the chickens to move from one living area to the next.

- The chicken coop is made out of waterproof fir wood and is insect- resistant as well, thus contributing to the structure's durability.

- The wood material thus makes the structure more durable.

- The chicken coop has more than one entry point, thus making accessibility and cleaning easier.

- The chicken coop is surrounded by steel wires, which secures the chickens from predators such as snakes.

- There is a run for the chickens to roam, but it is also removable, thus contributing to its flexibility.

- The chicken coop has a nesting box that has 2 divisions, thus giving 4 nesting spaces for the chickens to comfortably lay their eggs and the farmer to collect them easily.

- The chicken house is elevated but can be accessed through the ramp, and this protects the chickens from underground predators that could dig underneath, especially at night.

This product is available on Amazon.

28. PawHut 83" Wooden Backyard Chicken Coop with Covered Run and Nesting Box

The PawHut 83" Wooden Backyard Chicken with Covered Run and Nesting Box has the following features:

- The structure is made out of fir wood, thus making it water-resistant and protected from rotting, thus promoting its durability.

- The structure is spacious and has indoor space for the chicks to stay whenever it is too hot or the outside area where they seek to play in the sun.

- This outdoor and indoor are also promoting the ventilation within the chicken coop, and whenever the weather is cold, there is the option of locking up the accessibility.

- This chicken coop has a hinged nesting box for the chickens to lay their eggs and the farmer to collect the eggs easily.

- There structure also has perches inside the chicken coop for roosting.

- There is an outdoor run for the chickens to move around. The run is secured to protect the chickens from predators.

- The chicken coop does not take up a lot of space.

- The roof of the chicken coop can be opened, thus making it easier to access different areas when cleaning.

This product is available on Amazon.

29. Pets Imperial Dorchester Chicken Coop

The most distinct characteristic of this chicken coop is that it is a run that takes the shape of an arc. The run provides the chickens room to play and move around, thus contributing to their overall happiness.

The features of this chicken coop are as follows:

- The roof of this chicken coop can be opened, thus making cleaning easier since one can access the chicken coop from above. The opening of the roof also contributes to air circulation within the chicken coop, thus ensuring that there is no suffocation.

- The chicken coop does not take up a lot of space so it would be perfect for the backyard

- The chicken coop is made out of treated timber, thus making it last a long time.

- The chicken coop has a galvanized metal tray to collect the chicken droppings, thus ensuring the hygienic standards of the coop.

- The coop can only accommodate 2 to 3 chickens and would, therefore, be ideal for a beginner.

- The coop is easy to assemble because there is a detailed instruction manual.

This product is available on Amazon.

30. PawHut Wooden Bunny

This chicken coop is multi-purpose, meaning that it can accommodate a variety of animals such as rabbits, chickens, guinea pigs, among others.

The features of this chicken coop are as follows:

- The structure does not take up a lot of space.

- The chicken coop is made out of fir wood, thus making it water-resistant and contributing to its durability.

- The structure is firm, thus protecting the chickens from all kinds of weather conditions and predators such as foxes.

- The chicken coop has a spacious outdoor run for the chickens to move around.

31. Lazymoon Wooden Chicken Coop

The best feature of this chicken coop is that it is made out of fir lumber. This means the structure will be durable because fir wood doesn't rot and is protected from damage by insects.

Other features of this chicken coop are as follows:

- There is an enclosed outdoor run for the chickens to move around and enjoy the sunlight, but it is also secure enough to protect them from predators.

- There is a nesting box for the chickens to comfortably lay their eggs and the farmer to collect them easily.

- The chicken coop has a tray to pull out when collecting the chicken droppings, thus promoting cleanliness within the chicken coop.

- This chicken coop can only accommodate 2 to 3 chickens.

This chicken coop is available on Amazon.

32. Cocoon Hatched Chicken Coop with Night Shutter

This chicken coop doesn't take up much space, and it is very attractive

This chicken coop has the following features:

- The chicken comes has a run for the chickens to move around and play thus making them comfortable

- The coop is compact as it rests on 220 x 75 x 113cm worth of space.

- There is a door that distinguishes the run from the nesting and roosting area, which increases the security of the chickens from predators, especially at night. The door is also helpful during harsh weather to ensure that the chickens are warm, especially the chicks.

- The chicken coop is made of high quality, thus ensuring that it is a strong structure.

- There is a nesting box for the chickens to comfortably lay their eggs and the farmer to collect them easily.

- There are perches for the chickens to roost on.

- The nesting box and perches are removable which would be conducive for the farmer when cleaning and where there is a need to create more room for the chickens to play

- The floor can be removed, thus making it easier for the farmer to clean the coop.

- The chicken coop would be ideal for a beginning poultry farmer because it can only hold 2 to 3 chickens.

The challenges with this chicken coop are as follows:

- It is relatively expensive.

- The run is not big enough, and there may be a need to open the chicken coop so that the chickens can roam outside the coop, which is not very secure since they could be open to predatorial attacks.

The product is available on Amazon.

33. FeelGoodUK Poultry Ark

This chicken coop design has been well thought out but can only accommodate 2- 3 chickens since it takes up 175 x 66 x 100cm worth of space, which is very minimal.

The features of this chicken coop are as follows:

- There is nesting for the chickens to lay their eggs and the farmer to easily collect the eggs.

- There is a perch for the chickens to rest on.

- The chicken coop is covered, thus protecting the chickens from rain.

- The size of the chicken coop is minimal, but the floor is large enough for the chickens to move around comfortably.

- The coop is long-lasting because it is made out of twin-insulated plastic wall panels and wood.

- There is a metallic tray that would be collecting the chicken droppings that the farmer would be pulling out when cleaning.

The demerits of this chicken coop are as follows:

- The material used in making the structure makes it more durable it also makes it very light that animals could attack it easily and have access to the chickens.

- The structure doesn't have a run for the chickens to move around; therefore, a farmer may need to purchase or build one.

This product is available on Amazon.

34. FeelGoodUK Large Chicken Coop House

This chicken coop can only accommodate 4 chickens measuring at 85 x 115 x 90cm, making very compact. However, the Royal Security for the Prevention of Animals states that this chicken coop could accommodate 8 chickens. The 4 chickens would be ensuring that the chickens are comfortable.

The features of this chicken coop are as follows:

- The chicken coop is made out of fir wood, which is stuck together using glue rather than nails, thus making it insect and waterproof.

- There is a nesting box with 2 nesting spaces for the chickens to lay their eggs and the farmer to collect them easily.

- There are 3 perches for the chickens to roost on.

- The hinged doors have locks so that to add more security for the chickens against predators and cold weather conditions. The opening of the front and back doors also contributes to the ventilation of the chicken coop.

- There is also a galvanized steel tray to collect the chicken droppings, thus making cleaning easier.

The challenges with this chicken coop are as follows:

- Despite the structure being made from fir wood, the wood used in this structure is thin, which could, in turn, affect its durability.

- The chicken coop doesn't come with an attached run; therefore, a farmer may need to purchase a separate run or build one, which would be an additional cost.

This product is available on Amazon.

35. Cocoon Hen House

This chicken coop can house 4 large chickens or 6 bantam chickens because of the structure measures at 154 x 66 x 110cm.

This chicken coop has the following features:

- This chicken coop has 2 spacious nesting boxes that have been divided into 2 for the chickens to comfortably lay their eggs and the farmer to be able to collect them easily.

- There are locks on 2 doors to protect the chickens from predators, especially at night, and the locks can help during the cold weather as well. Having a front and rear door also improves ventilation within the chicken coop as well.

- The chicken coop also has a metallic tray for the collection of chicken dropping, which the farmer can pull out and clean.

The demerits of this chicken coop are as follows:

- The chicken coop is made out of timber, but the wood has not been treated; therefore, it is not yet water-resistant, and a farmer would need to treat the wood themselves.

- The instruction manual has not been well written out for the farmer to assemble easily

Building a Chicken Coop

Despite there being advantages to purchasing a chicken coop, building one also has its advantages. The following are a few advantages of building a chicken coop.

- It can be less expensive

Sometimes someone already has the raw materials needed to build a chicken coop lying around their house, and they only need to purchase a few more things. That being the case, it would be cheaper to build than to purchase a chicken coop. There is also the possibility of repurposing unused structures into chicken coops such as a playhouse for children or even a kennel.

- One could build a fully loaded chicken coop

Some chicken coop designs are available for purchase but lack certain things that are very crucial for the overall happiness of the chickens. Some designs do not have a run for the chickens to roam in, forcing a purchaser to build or purchase a separate run for the chickens. Some designs lack nesting boxes for the chickens to lay their eggs, thus making it harder for a poultry farmer to collect the eggs. Some designs lack perches for the chickens to roost, thus forcing the chickens to sleep on the floor of the chicken coop, which is unnatural for them. Perches are important for chickens because, in the wild, chickens would roost on higher ground as a means of protecting themselves from predators. There are also some chicken coop designs available in the market that lack a ventilation system that can be regulated for the hot and cold weather.

- The builder has the option of choosing high-quality material for the chicken coop

As earlier discussed, the durability of the chicken coop solely depends on the material used in the building. Chicken coops can be made out of wood, plastic, and metal, but wood has been proven to be the most preferable. However, not all wood guarantees durability. A builder could have the money to purchase old heartwood cedar, which is a hardwood that is rot-resistant but expensive. The building of the chicken coop opens one up to more options.

- The ability to add extra features

The building of the chicken coop would allow someone to add certain features that are not available in most bought chicken coop designs. An example would be the addition of a garden planter to grow flowers or a few vegetables. The chicken droppings could also be preserved to be used as manure for the plants. Another example would be the addition of heating lamps to warm the chickens during the cold weather, especially when they are chicks.

- One has the option of choosing the size and shape of the chicken coop that they want

There are very many chicken coops sizes and shapes in the market, but maybe they do not satisfy all the needs of the purchaser. A chicken farmer may want a chicken coop that is large enough for them to walk inside, thus making it easier to clean but also include a bench inside the chicken coop run so that they can watch the chickens play and also pet them. When building a chicken coop, they would have the option to choose the size and shape that they would prefer.

- A builder would have the option of choosing the exact color and design of the chicken coop he/she wants

A builder may want a chicken coop design and color that would fit well with the aesthetic of the house. This would be to ensure that the chicken coop blends well in the backyard of the house, thus making it even more appealing.

- An opportunity to bond with the family

The building process of the chicken coop doesn't have to be a stressful time because it can be used as a time where the family gets to bond with one another. A parent could also teach a child how to build structures, which would be a learning moment for him/her. Some people also find the idea of building something of their own fulfilling.

Steps to follows when designing a chicken coop
- Identifying the space that the chicken coop will take up.

Go to your backyard and decide on whether you want a large, medium, or small chicken coop. The size available or that which you are willing to use will be a huge factor because it will determine the size and design of the chicken coop. Space will also give one an idea of the maximum number of chickens they can rear depending on the size of the breed. The recommended space requirements are as follows.

i. The floor space - 2 square feet for each chicken to ensure that there is no overcrowding within the chicken coop. Overcrowding could lead to easy transmission of diseases among the flock, suffocation due to insufficient oxygen and fighting, which would lead to injuries on the chickens. Where the breed is bigger than the size of a standard chicken, it is recommended that one should one square foot, thus making it 3 squared foot.

ii. A nesting box should be available for 2-3 chickens, which will create a conducive environment for egg-laying such as privacy and darkness. The nesting box will also make it easier to collect the eggs.

- Sketching of the chicken coop

One of the most important things to do would be to research the multiple chicken coop designs that are out there as a source of inspiration.

Upon identifying the chicken coop design, identify whether you intend to build the chicken coop as it is or add a few modifications which will lead us to our first step, which is sketching.

This step would also include the builder to calculate the measurements needed to build the chicken coop based on the design selected.

- Space, where the chicken coop shall be built, should be marked.

This area selected should also consider whether there will be sunlight accessibility for the warmth of the chickens and where the wind blows, especially on a hillside. The marked ground should include the space for a chicken run so that chickens can roam in. Whether the chicken runs should be open or not will depend on the design or the availability of predators. The structure to be built should be elevated to protect the chickens from predators that could dig underneath to access the chickens and to ensure that water doesn't push the chicken coop, especially during heavy rainfall.

- The following step would be to identify the features that the chicken coop ought to have which are as follows:

    i.   A nesting box for the chickens to lay their eggs.

    ii. A floor within the chicken coop for the chickens to walk on. It could be wooden on concrete, depending on the available time and money.

    iii. A door for the chickens to access and to increase ventilation within the chicken coop.

    iv. A ramp for the chickens to walk on to access the outside from the door.

    v.   A removable roof so that it would be easier to clean the inside of the chicken coop.

vi.   A removable tray that would be collecting the chicken droppings, thus making it easier to clean.

vii.   Perches for the chickens to roost on.

viii.   Electricity, if need be, especially during the cold season. One could connect heating lamps to ensure that the chicken coop is warm. Low light bulbs would deceive the chickens that it is daytime, thus ensuring and increasing their egg production rate.

ix.   The items to use to feed and supply water to the chickens such as feeding and drinking trough.

x.   The chicken padding material that will be inside the nesting box to cushion the eggs that will be laid.

- The next step would be selecting the chicken coop materials that would be best to ensure that the structure will be strong and long-lasting, which would be wood as earlier discussed.

Other materials would include fasteners such as nails and screws, among others. The necessary tools needed to build the chicken coop would also be crucial to identify at this stage.

- The next step would be to consider the chicken coop ventilation system that would be suitable for its location.

36. The A-Frame

This chicken coop is also referred to as a Tractor Coop. There are numerous A-shaped chicken coops, but this is just but one. The coop measures at a measurement of 7 x 6 feet.

The features of this chicken coop are as follows:

- The structure is made out of part wood, thus benefiting from the sturdy characteristics of wood. This also means that the chicken coop will be durable and will not cost the owner the financial cost of replacing it. The wood will also be able to withstand harsh climatic conditions.

- There is a metallic roof that will protect the chickens from rain and make the room warmer for them.

- The chicken coop is easy to assemble and therefore, would be great for a person who would be rearing chickens for the first time.

- The chicken coop is also relatively cheap.

- The chicken coop is portable thus can be moved around at the convenience of the farmer, especially where the farmer intends to use the chicken, dropping as manure within the area where the chicken coop is located. The movement also reduces intense scratching by the chickens on one specific area.

- The structure can only accommodate 5-6 chicken.

- The materials needed in setting up the structure would be wood, nails, hinges, and door pulls, wire mesh, and galvanized roofing.

- There is ample ventilation because the lower part of the chicken coop is made out of mesh.

- The chickens can access the floor where they could eat all kinds of bugs such as worms which would be good for their nutrition

- The bottom floor is open to the ground, and such doesn't have to be cleaned.

- The upper part of the chicken coop makes it conducive for the chickens to rest during the cold weather.

- Due to the ability to customize it, a farmer could include heat lamps to warm up the chicken coop during the cold season. This would be extremely beneficial where there are chicks in the coop.

The demerits of this chicken coop are as follows:

- There is no elevation for the chickens to protect the chickens from predators that could dig underneath the boundary to access the chickens. As a means of deterring this, a farmer could opt to attach mesh at the bottom of the chicken coop, but the challenge with this is that this could scratch on the land when attempting to move the chicken coop. This would, therefore, only work if one lets go of this portable feature.

- An aggressive chicken would be hard to separate from the other chickens, and as such, some of the chickens could get hurt in the process, which would affect the farmer's productivity.

37. The Wheeled Chicken Coop

This chicken coop has wheels at the bottom, thus making it portable. This is conducive for a farmer where the intention to move it around so that the chickens can access different vegetation. The other benefit is that because chickens are always scratching the floor, this limits excessive scratching in one area.

The construction of this chicken coop does not involve a run, and such one can modify this chicken coop to include a run for the chickens to move around. This chicken coop is small and can only accommodate six to 8 chickens. One could also modify the chicken coop to include the A-frame design to accommodate more chickens. This will reduce overcrowding within the chicken coop.

The advantages of this chicken coop are as follows:

- The mobility of the chicken coop allows the chickens to diversify their diet because they could eat the insects on the ground when they are scratching, such as worms.

- The chicken coop guarantees safety to the chickens from predators because it is enclosed, and upon a custom run being included, the chickens can comfortably play without fear of being attacked. The chicken coop is also elevated because the chicken coop stands above the wheels, which prevents underground predators from accessing the chickens.

- There is a door for the chickens to access the outside and to secure the chickens at night.

- Where the owners of the chickens intend to relocate, they can easily move with the chicken coop because it is portable.

The demerits of this chicken coop are as follows:

- The chicken coop cannot accommodate many chickens. Therefore, it would not be lucrative where an owner intends to make a profit from poultry farming.

- The chicken coop is too small for a human being to walk inside since it is even small enough to be moved.

- The chicken coop is not spacious inside, and there is the risk that there might be overcrowding within the chicken coop.

- The chicken coop doesn't have a large entry and many access points; therefore, it would be difficult to clean and collect eggs.

This chicken coop would be ideal for a family as a hobby.

38. Traditional Chicken Coop

This chicken coop can only accommodate 4 to 6 chickens. The features of this chicken coop are as follows:

- There is a nesting box for the chickens to lay their eggs comfortably. The availability of nesting boxes will make it easier for a farmer to collect the eggs and also protect the eggs since they are cushioned on material such as dry grass.

- There is a door so that the chickens can access the outside and also enclose them at night to protect them from predators and keep the chicken coop warm.

- One could also custom build a run for the chickens on moving around, thus promoting animal welfare. Where a run is included, this design has a ladder for the chickens to go and down to access the run.

- This design involves heat lamps to warm the chicken coop, especially during the cold weather. This is also important where there are chicks because they could easily die from the cold.

- In comparison to the A-frame design, this chicken coop is more spacious, thus giving the chickens more room to move around. Space is very important because it ensures that the chickens, especially during feeding, do not jump on each other, thus suffocating the weaker chickens. Space will also be useful because it will reduce the spread of diseases within the chicken coop.

The disadvantages of this chicken coop are as follows:

- This is a permanent structure, and, in most cases, permanent structures have a stronger odor, particularly during the hot weather. This means that the chicken coop will have to be cleaned more often than a portable structure.

- This chicken coop design is more complicated to build and expensive than the A-frame chicken coop design. Another option would be purchasing the design, but it would be more expensive; therefore, it would be better to assemble the raw materials and build one.

- This chicken coop design is enclosed, and therefore it would be very important to include a run for the chickens to roam or to allow them to move around within the compound. Failure to which the chickens will not be happy, which will, in turn, affect their productivity.

39. The Walk-In Hen House Chicken Coop Design

As the name suggests, this chicken coop design allows for a person to walk inside. The advantage of this chicken coop design is as follows:

- Since the chicken coop can accommodate a person; it would be easier for someone to clean because they can easily access the chicken coop.

- This design is very spacious for the chickens to play and socialize, thus promoting their overall happiness. Space also prevents overcrowding, which can irritate the chicken, leading to fights amongst each other. These fights could cause injuries that would open doors to infections, and at the end of the day, the production date will decrease.

- This chicken coop can accommodate ten to fifteen chickens because it is large.

- In as much nesting boxes are for the chickens to lay their eggs, sometimes the chickens lay their eggs elsewhere within the chicken coop. The great thing about this design is that one could easily inside and see the eggs that are not inside the nesting boxes and collect them.

- There is plenty of air circulation within the chicken coop due to the availability of the mesh for ventilation.

- This design allows for electricity to be allowed into the chicken coop. This would be an added advantage, especially where one would wish to access the chicken coop at night. Another would be that one could add heating lamps when the weather is very cold.

The disadvantages of this chicken coop are as follows:

- This design is very expensive; therefore, many intending farmers do not, but.

- The design is also difficult to build in comparison to portable and smaller chicken coop designs. It requires certain skills to build, which is a limit for some people. One would also need to purchase or own certain tools to build it, such as saws to cut the wood, among others.

- The chicken coop floor will need a proper foundation for it to be sturdy; therefore, one needs a concrete floor. This would be an additional cost to set up the chicken coop

- Where the owner of the chicken coop intends to relocate, it would not be possible for them to do so with the chicken coop because it is quite large.

- Due to the absence of removable trays, the chicken droppings will be all over the floor of the chicken, thus making cleaning the chicken coop a task.

40. The Ritz

This chicken coop design is also referred to as the large hen house. The features of this chicken coop are as follows:

- The chicken coop can accommodate a large number of chickens of up to 50 chickens.

- This design has a vent to facilitate ventilation within the structure. The aeration is also increased with the availability of windows and a mesh.

The advantages of this chicken coop are as follows:

- The chicken design is also walk-in, and as such, a person could enter and easily clean the chicken coop. It would, therefore, be much easier to maintain a high hygienic standard with the chicken coop.

- This structure is long-lasting and can withstand harsh weather.

- This chicken coop design allows for someone to do poultry farming on a commercial basis because it can accommodate many chickens.

- This design is very spacious for the chickens to play and socialize, thus promoting their overall happiness. Space also prevents overcrowding, which can irritate the chicken, leading to fights amongst each other. These fights could cause injuries that would open doors to infections, and at the end of the day, the production date will decrease.

- This design allows for the installation of electricity within the chicken coop. This would be an added advantage, especially where one would wish to access the chicken coop at night. Another would be that one could add heating lamps when the weather is very cold.

- This chicken coop can accommodate ten to fifteen chickens because it is large.

- In as much nesting boxes are for the chickens to lay their eggs, sometimes the chickens lay their eggs elsewhere within the chicken coop. The great thing about this design is that one could easily inside and see the eggs that are not inside the nesting boxes and collect them.

The disadvantages of this chicken coop are as follows:

- The design is also difficult to build in comparison to portable and smaller chicken coop designs. It requires certain skills to build, which is a limit for some people. One would also need to purchase or own certain tools to build it, such as saws to cut the wood, among others.

- The chicken coop floor will need a proper foundation for it to be sturdy; therefore, one needs a concrete floor. However, one could also sand on the floor.

- This chicken coop design does not have removable trays, and as such, the chickens will excrete all over the floor. The chicken's droppings will be all over the floor of the chicken, thus making cleaning the chicken coop a task.

- This design is very expensive because it requires a lot of material to set up and therefore may not be able to afford it

- Where the owner of the chicken coop intends to move from their house, it would not be possible for them to do so with the chicken coop is a permanent structure. Therefore, before building this chicken coop, one would have to establish whether they will be relocating or not.

41. A Playhouse Chicken Coop

This chicken created from a children's playhouse. The structure is ideally made out of wood, thereby making it sturdy. However, the durability of the chicken coop will solely depend on the kind of wood used. The features of the chicken coop will solely depend on the kind of playhouse repurposed. There was a person who created a chicken coop out of the moderately sized playhouse, and that shall be our specifications on the features.

The features of the chicken coop are as follows:

- This chicken coop can accommodate nearly six to nine chickens. Therefore, it would be more conducive for a beginner.

- The chicken coop is composed of 2 or 3 nesting boxes for the chickens to lay their eggs comfortably. The nesting boxes provide sufficient conditions for the chickens to lay their eggs, such as privacy and darkness.

- The floor of the chicken coop was built to be easily removable to collect the chicken droppings, thereby ensuring that the chicken coop is clean.

- There are 2 windows on the structure to allow ventilation, thus ensuring that the lives of the chickens are sustained.

- The chicken coop is raised from the ground. This ensures that predators cannot access the chickens from the ground.

- The chicken coop has a ramp for the chickens to walk on to access the outside.

- The chicken coop has a door which contributes to the ventilation of the chicken coop.

The disadvantages of this chicken coop are as follows:

- The absence of a run prevents the chickens from safely moving around. This would force the owner to build or buy a run because allowing the chickens to roam around without enclosure could be opening them to predatorial attacks

- The chicken coop is not spacious and would, therefore, be too crowded for the chickens. This later becomes a health hazard because chickens could fight and easily harm each other.

42. Backyard Chicken Barn

The appearance of a barn inspires this chicken coop design. The design has the following features:

- Based on its barn inspiration, it is reddish in color

- This design has a run for the chickens to roam. The run is enclosed to secure the chickens from predators.

- The chicken coop design is elevated, thus protecting the chickens from predators that could dig underground to access the chickens.

- The chicken coop has sufficient ventilation due to the presence of windows. The top of the chicken coop also has a chimney-like structure to give room for the warm air within the chicken coop to rise.

- The chicken coop is made out of wood, thus making it a sturdy structure.

- When building the chicken coop, one should also include nesting boxes for the chickens to lay their eggs, which will also make it easier for the owner to collect the eggs.

- The chicken coop does not require expensive material to build. However, ensure that the wood used isn't pressure-treated lumber because the preservative used is harmful to chickens.

The disadvantages of this chicken coop designs are as follows:

- The chicken coop design can only accommodate 3 to 5 chickens

43. The Big and Cheap Chicken Coop

This chicken coop is not well built. It lacks essential features for the comfort and happiness of chickens. The features of this chicken coop are as follows:

- The chicken coop is made out of plywood.

- The chicken doesn't need a lot of materials and tools to build. The chicken coop tools are also easy to use and are available; for example, one needn't use power tools.

- The materials needed to construct are nails, screws, 2 hinges that will be used to secure the door and wood.

- The chicken coop is elevated, thus protecting the chickens from predators that could dig and have access to the chickens.

The disadvantages of this chicken coop are as follows:

- This chicken coop lacks a run for the chickens to move around. This would mean that a farmer needs to purchase or build a run. The other option would be to allow the chickens to roam around in the open space, but animals could attack the chickens.

- The design does not have nesting boxes for the chickens to lay their eggs. The absence of the egg boxes discourages the chickens from laying the eggs. The eggs that are laid inside the chicken coop could easily crack or break, thus exposing them to contamination.

- The chicken coop doesn't have perches inside for the chickens to roost. The chickens are forced to sleep on the floor of the chicken coop.

- The chicken coop is small and can, therefore, only accommodate a few chickens.

- The chicken coop does not have a ramp for the chickens to easily access the outside.

44. The Lollipop Chicken Coop Design

This chicken coop design is shaped like a popsicle. The features of this chicken coop design are as follows:

- The chicken coop has a front and back window made out of wire mesh for proper aeration of the chicken coop. This ensures that the chicken coop has sufficient ventilation.

- The chicken coop design has a staircase for the chickens to easily go up and down to access the outside and the inside of the chicken coop.

- This design has 2 nesting boxes so that the chickens feel comfortable enough to lay their eggs, thus increasing productivity.

- The chicken coop is made out of wood while the floor is specifically made out of plywood.

- The chicken coop is elevated to protect the chicken from attacks from predators.

The disadvantage of this chicken coop is that the design does not include a run for the chickens to roam around. A farmer could, therefore, redesign the structure to include a run for the chickens to roam within safely.

45. Doghouse Chicken Coop Design

This chicken coop design is built from a previous dog house. The features of this chicken coop design are as follows:

- The chicken coop has a ramp for the chickens to access the outside easily

- The chicken coop is made out of wood. The builder of the chicken wood should select wood that is safe for chickens, for example, preserved softwood, which will ensure that it is waterproof and insect-resistant, which will, in turn, ensure the durability of the chicken coop. Wood, such as pressure-treated lumber, would be hazardous to the chickens because chickens tend to scratch the floor, and they could end up being exposed to the chemical preservatives.

- The chicken coop only has one access point, which is the door.

The advantages of this chicken coop are as follows:

- The chicken coop is not expensive to build because it doesn't need expensive materials.

- The chicken coop is easy to create from an already available dog house.

- The chicken coop design is elevated, thus protecting the chickens from predators that could dig underneath to access the chickens.

The disadvantages of this chicken coop design are as follows:

- This chicken coop design can only accommodate 2 to 4 chickens

- The design does not include a run for the chickens to roam in. The owner would have to custom build or purchase a run so that the chickens are protected from predators when they move around.

- The design chicken coop is not spacious enough for the chickens, which could cause them to fight each other.

- The dog house design does not include nesting boxes for the chickens to comfortably lay their eggs. The chicken coop is small; therefore, a farmer would still easily collect the eggs, but since the eggs do not have a cushioning, they could easily crack or break.

- The chicken coop doesn't have windows, and therefore there isn't enough ventilation when the door of the chicken coop is closed.

46. Scandinavian Style Chicken Coop Design

This design is very stylish and would look incredible in your backyard. The structure was built and designed by David Manchester.

The features of this chicken coop are as follows:

- The chicken coop is elevated from the ground to protect the chickens from underground predatorial attacks.

- The chicken coop has run for the chickens to move around in. The run is also enclosed to protect the chickens from predators.

- The chicken coop is spacious for the chickens to roam in comfortably. This ensures that there is no overcrowding within the chicken coop, thus ensuring the comfort of the chickens.

- The design is large enough for a human being to access. This makes it much easier to clean the chicken coop.

- The chicken coop has nesting boxes for the chickens to lay their eggs comfortably.

- The design has light fixtures outside the chicken coop, thus making it easier for a human being to access at night whenever necessary.

- The walk-in door has a glass window on it to allow for light to get inside the chicken coop during the day, thus contributing to the egg-laying process and the absorption of Vitamin D by the chickens.

- There is a small door for the chickens to access the enclosed chicken coop. A ramp is also attached to this door for the chickens to easily move from the chicken coop to the run and vice versa.

- The chicken coop is made out of wood, thus making it a sturdy structure.

The disadvantages of this chicken coop design are as follows:

- The chicken coop needs adequate space to build on.

- The design is a permanent structure; therefore, it should only be built on an area where the owners don't intend to relocate

- The design is expensive to build.

47. Hillside Chicken Coop Design

This chicken coop design is a smart way to place a chicken coop on a hill. The features of this chicken coop are as follows:

- The chicken coop is large enough to allow a person to walk inside. This would make it easier for a person to clean the inside of the chicken coop.

- The chicken coop is large enough to be confused for a small house.

- The chicken coop has a large enclosed run for the chickens to move around. The enclosure of the chicken run protects the chickens from predators during the day. The run also has a door on the run that a human being can fit through.

- The chicken coop has many windows to allow for ample ventilation within the chicken coop, thus sustaining the lives of the chickens.

- The chicken coop design is elevated from the ground to protect the chickens from animals that could dig underneath the chicken coop to access the chickens inside the coop.

- There are a small door and ramp for the chickens to be able to move to and from the chicken coop and run.

- The structure is made out of wood, thus making it sturdy.

- The chicken coop is spacious for the chickens to fit comfortably.

- When building the chicken coop, one should also include nesting boxes for the chickens to lay their eggs and perches for them to roost.

The disadvantages of this chicken coop are as follows:

- The chicken coop is expensive to build.

- The design requires a certain amount of skill to construct, especially because it is in an inclined position.

48. Farmhouse Chicken Coop Design

Mikee Krieg created this chicken coop design. The features of this chicken coop are a follow:

- This chicken coop design has a weather vane. The purpose of this is to establish the amount of wind blowing to regulate the ventilation within the chicken coop.

- The chicken coop is made out of wood to ensure that the structure is long-lasting.

- The chicken coop is also beautiful with flower baskets dangling from it.

- The chicken coop is elevated to protects the chickens from predatorial attacks, especially at night.

- The design comes with an attached chicken run for the chickens to move in. The run is enclosed to protect the chicken from predatorial attacks during the day.

- The design has transparent windows on the side for sunlight and can be opened for aeration.

- Since this is a custom design, one should include nesting boxes for ease of collecting the eggs and perches for roosting.

- The chicken coop is also spacious.

The disadvantages of this chicken coop design are as follows:

- The chicken coop is expensive to build.

- The chicken coop also takes up a lot of space.

- It is a permanent structure; therefore, before building this chicken coop, one should first know whether they would be relocating.

49. The Mini Chicken Coop Design

This chicken coop is ideal for a farmer that intends to keep very few chickens, for example, two or three. The features of this chicken coop can be built to include:

- A nesting box to create the right condition for the chickens to lay their eggs. The availability of a nesting box would also make it easier for a farmer to collect the egg.

- The design can also be elevated to prevent predators from digging underneath the chicken coop to access the chickens.

- The chicken coop could also include a small enclosed run for the chickens to move around safely.

- One could also decide to make the chicken coop portable or permanent.

- The chicken coop takes up very little space.

- The chicken coop is not expensive nor complicated to build.

The disadvantages of this chicken coop are as follows:

- The chicken coop can only accommodate a few chickens.

50. The Ranch Style Chicken Coop Design

The design of a ranch inspires this chicken coop design. The features of this chicken coop design are as follows:

- The chicken coop can accommodate many chickens because it is large.

- The design is spacious for the chickens to move around and socialize comfortably.

- There is an enclosed run for the chickens to move around during the day safely.

- The chicken coop has nesting areas to facilitate egg-laying. The availability of nesting boxes makes it easier for someone to collect the eggs.

- The structure has multiple windows for ample ventilation within the chicken coop.

- The chicken coop has a human-sized door for someone to access. This makes it easier to clean the chicken coop.

- There is a ramp between the chicken coop and the run so that the chickens can access either side.

The disadvantages of this chicken coop are as follows:

- The chicken coop needs a large space to build on.

The design doesn't have an elevation, which means that predators can dig underneath to access the chickens on the other side.

Conclusion

In conclusion, there are very many chicken coop designs that are available in the market for a farmer to choose from. Therefore, the perfect chicken coop will solely depend on the needs of the poultry farmer. However, there are certain features that we have established that would be very crucial for the chicken feed to have, such as the ease of cleaning it, protection from external harm such as the predators, and the rain, among others. Therefore, those basic requirements can be the ground basis when shopping for a chicken coop. Where a farmer intends to housing a large flock, they should be aware that the bigger the chicken coop, the higher the financial cost. A person intending to purchase a chicken coop will have to consult and do research on the perfect design that meets their demands.

References

Achilli A, Bonfiglio S, Olivieri A, Malusa A, Pala M, Kashani BH (2009) "The multifaceted origin of taurine cattle reflected by the mitochondrial genome" < https://www.nature.com/articles/hdy201283 citeas

Naked Chicken Council 'U.S Chicken Industry History '(Nation Chicken Council)< https://www.nationalchickencouncil.org/about-the-industry/history

Mench, Joy & Sumner, Daniel & Rosen-Molina, J. "Sustainability of egg production in the United States-The policy and market context (2011) < https://www.researchgate.net/publication/49701681_Sustainability_of_egg_production_in_the_United_States-The_policy_and_market_context/citation/download

And Schneider, Six Benefits of Owning Chickens' (Acreagelife, July 26th 2018) < https://www.acreagelife.com/hobby-farming/six-benefits-of-owning-chickens

Linley Tulloch,'Animal Rights and the Torturous Suffocation of Chickens' (top scoops, 7th December 2019) < https://www.acreagelife.com/hobby-farming/six-benefits-of-owning-chickens

Dowarah, Runjun, "The Role Of Poultry Meat And Eggs In Human Nutrition" (2013)

<https://www.researchgate.net/publication/321254565_THE
_ROLE_OF_POULTRY_MEAT_AND_EGGS_IN_HUMA
N_NUTRITION

Lesa (Better hens and gardens of Bramblestone Farm) "Chicken
Coop Designs-10 Important features" <
https://www.betterhensandgardens.com/chicken-coop-
designs-10-important-features

Erin Phillips "6 Basics for Chicken Coop Design" (Backyard
Poultry, May 2nd
2020)<https://backyardpoultry.iamcountryside.com/coops/b
asics-chicken-coop-design

Paige Cerulli, "Chicken Nesting Boxes DIY, Care and Why they
are Important" (Wide Open Pets) <
https://www.wideopenpets.com/chicken-nesting-boxes-diy-
care-and-why-
theyreimportant/:~:text=Why%20Nesting%20Boxes%20Ar
e%20Important&text=Nesting%20boxes%20are%20essenti
al%20supplies,yard%20or%20in%20your%20coop

Jennifer Poindexter "How to Clean Your Chicken Coop & Run: 9
Tips to do it Right" (Morning Chores) <
https://morningchores.com/cleaning-chicken-coop

"5 Ways to Make Coop-Cleaning Easier" (Hobby Farms) <
https://www.hobbyfarms.com/5-ways-to-make-coop-
cleaning-easier-
3/:~:text=To%20clean%20the%20coop%2C%20we,1%20h
our%20to%20re%2Dbed.

Technical Team, Ziggity Systems, Inc, "The role of water in the life
of a chicken" < https://ziggity.com/wp-
content/uploads/helpfularticles/cont_article_pdf_20.pdf

Patrick, "Should you put food and water in the chicken coop"
        (Silkie, May 17th 2020) < https://silkie.org/food-and-water-
        in-the-chicken-
        coop.html:~:text=Reasons%20not%20to%20put%20food%
        20and%20water%20in%20the%20chicken%20coop%3A&t
        ext=Feed%20can%20spoil%20and%20grow,if%20the%20fl
        ock%20is%20indoors

'Chicken Coops' (PoultryHub)
        http://www.poultryhub.org/production/backyard-village-
        poultry/raising-backyard-poultry/chicken-coops

Kassandra Smith "Top 20 Chicken Breeds for your Backyard"
        (Backyard Chicken Coops, 26th June 2020)<
        https://www.backyardchickencoops.com.au/blogs/learning-
        centre/top-20-chicken-breeds-for-your-backyard-coop

Chris Harris, "Key Factor for Poultry House Ventilation" (The
        Poultry Site, 25th January
        2012)<https://thepoultrysite.com/articles/key-factors-for-
        poultry-house
        ventilation:~:text=Ventilation%20systems%20are%20gener
        ally%20divided,mechanical%20air%20movement%20(fans)

TopBest10Reviews, Top 10 Best Chicken Coop Reviews/Buyer's
        Guide <https://www.topbest10reviews.com/best-chicken-
        coops

Coop Thoughts, 'Which wood is best for a chicken coop?'
        <https://www.thegardencoop.com/blog/2009/02/06/which-
        wood-is-best-for-a-chicken-coop

TopBest10Reviews, Top 10 Best Chicken Coop Reviews/Buyer's
        Guide <https://www.topbest10reviews.com/best-chicken-
        coops

Kassandra Smith "Top 10 Reasons to buy NOT build your chicken coop" (Backyard Chicken Coops, 10th December 2018)< https://www.backyardchickencoops.com.au/blogs/learning-centre/top-10-reasons-to-buy-not-build-your-chicken-coop

Erica Puisis 'The 6 Best Chicken Coops of 2020' (The Spruce, 27th April 2020) < https://www.thespruce.com/best-chicken-coops-4153598

'Bet Chicken Coops of 2020 Reviews' (Gardening/ Generators/ Backyard Chickens/ Hydroponics) <https://rurallivingtoday.com/backyard-chickens-roosters/best-chicken-coops

' Top 10 Best Chicken Coops 2020 Reviews, Buyer's Guide' (ctopreviews) <https://www.ctopreviews.com/best-chicken-coops

Candace Osmond, 'The Best Chicken Coops for the Backyard'(Backyard Boss) <https://www.backyardboss.net/best-chicken-coop-reviews

Pansy, 'Top 10 Best Chicken Coops 2020 Review'(Review Best 1, 20th of April 2020) < https://reviewbest1.com/best-chicken-coops

Sarah Bennet, 'Top 10 Best Chicken Coop: Review & Guide 2020' (Home Pets,13th of August 2019) <https://happyhomepets.com/best-chicken-coop

'Top 10 Best Chicken Coop Reviews/Buyer's Guide' (TopBest10Reviews) <https://www.topbest10reviews.com/best-chicken-coops

April Foot 'UK' s Best Chicken Coops for Keeping Your Hens Safe' (UP Gardener) <https://upgardener.co.uk/best-chicken-coop

Community Chickens '5 Key Steps to Building a Chicken Coop'
(Community Chicks, 20th of March 2018)
<https://www.wideopenpets.com/5-key-steps-to-building-a-
chicken-coop

Dave Malcom, '5 Practical Chicken Coop Designs' (Community
Chickens, March 6th 2019) <
https://www.communitychickens.com/5-practical-chicken-
coop-designs

Andreanna Lefton 'The Most Beautiful Chicken Coops We have
Ever Seen' (Bob Villa) <
https://www.bobvila.com/slideshow/the-most-beautiful-
chicken-coops-we-ve-ever-seen-44309

www.ingramcontent.com/pod-product-compliance
Lightning Source LLC
Chambersburg PA
CBHW061303120726
48001CB00001B/449